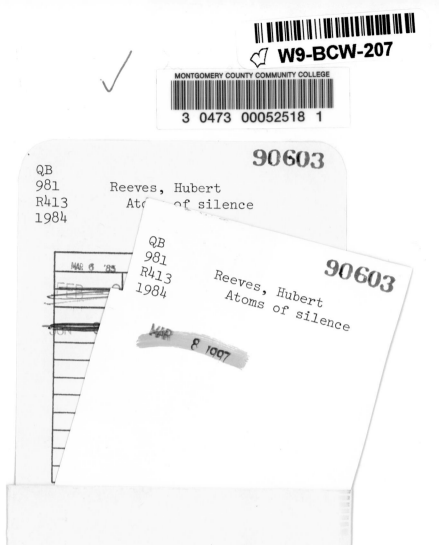

LEARNING RESOURCES CENTER
MONTGOMERY COUNTY COMMUNITY COLLEGE
BLUE BELL, PENNSYLVANIA

Atoms of Silence

Atoms of Silence

An Exploration of Cosmic Evolution

Hubert Reeves

translated by Ruth A. Lewis and John S. Lewis

The MIT Press
Cambridge, Massachusetts
London, England

Original edition © 1981 by Éditions du Seuil, Paris, France, published under the title *Patience dans l'azur: L'évolution cosmique.*

This book was set in Baskerville
by The MIT Press Computergraphics Department
and printed and bound by Halliday Lithograph
in the United States of America.

Library of Congress Cataloging in Publication Data

Reeves, Hubert.
  Atoms of silence.

  Translation of: Patience dans l'azur.
  Bibliography: p.
  Includes index.
  1. Cosmology. I. Title.
QB981.R413   1984   523.1   83-43019
  ISBN 0-262-18112-6

This book is dedicated to all who marvel at the universe.

Contents

It is often asserted that the scientific method is an impersonal, cold, matter-of-fact way to deal with our environment. Scientific research and its results are supposed to be far removed from the values, emotions, and sentiments that we associate with the word "human," even if the objects of the research are part of human life, as in biology or medicine. Natural science, it is claimed, has little to do with those experiences that are important in the world of feelings and emotions; science does not "turn us on," at least not those of us who are not directly involved in its activities.

The reasons for this assertion are manifold. Modern science does use concepts that are rather abstract and far removed from our daily experience. It also uses mathematics abundantly for the formulation of its ideas. But the most important reason is the lack of effort by the scientific community to tell the public, without mathematics, in a clear and easily understandable language what science is all about, how it works, how deeply involved it is in human concerns, how exciting it is, and how it can strengthen the bond between man and nature. Some scientists say that communicating at this level is impossible; others claim it is easy and proceed to publish so-called popular books transmitting scattered details that can be easily explained but do not add up to a unified whole. But then there are some, not many, who do succeed in giving readers an idea of the significance, breadth, and penetrating power of the great edifice of concepts called science.

The author of this book is one of them. Moreover he succeeds more than many others in impressing the reader with the unity of nature, how everything hangs together and is intimately connected, from the Big Bang to the living world, how the life blood of nature pulsates all through the evolution of matter from the hot gas at the start to the plenitude of materials and forms of life today. It is truly exciting reading, as it should be, since nothing, not the best science fiction, is as dramatic and overwhelming as the history of our universe. This history has slowly emerged before our intellectual eyes during the last few decades, but it is still partially shrouded by our ignorance and by our limited knowledge of what is going on in the depths of space and time. Nevertheless, what we know and infer today about this history is enough to make us humble and proud to be witnesses and actors in the great drama of existence.

The author of this book has found an adequate mixture of factual presentation and picturesque language to present natural science as it really is—a steady revelation to our mind of relations, connections, causes, and effects, demonstrating a unity in nature, a pulsating trend, a sense, and a direction. No longer do the immensity of space and time and the immutability of the laws of nature appear cold and inhuman. Rather, bonds are revealed in the drama of evolution that tie us all together—humans, animals, rocks, planets, stars, galaxies, and the immense expanses of space, full of all variety of forms, patterns, radiations, and forces. This book succeeds in showing that man and nature, the atoms and the stars, are all one great adventure, which we are here to behold and examine, and which we must protect lest it be destroyed by the misuse of the power we have gained from our deeper understanding of the universe.

What this book does and does not represent is well described by Walt Whitman's poem:

When I heard the learn'd astronomer,
When the proofs, the figures, were ranged in columns before me,
When I was shown the charts and diagrams, to add, divide, and
    measure them,
When I sitting heard the astronomer where he lectured with much
    applause in the lecture-room,
How soon unaccountable I became tired and sick,
Till rising and gliding out I wander'd off by myself,
In the mystical moist night-air, and from time to time,
Look'd up in perfect silence at the stars.

Hubert Reeves knows what the astronomer said; but he also is out there and looks up with Walt Whitman in silence at the stars.

Introduction: The Mountain and the Mouse

"Parturient montes, nascetur ridiculus mus."*
Horace, *Ars Poetica*

And the mountain brought forth a mouse. When you consider the relative sizes of the protagonists, the meaning of the ancient metaphor is unmistakable. It speaks of disproportion and disappointment, a lot of fuss and bother about nothing. And yet, when you consider relative complexity of organization, a rich paradox is revealed: With all its millions of tons of rock, a mountain can do nothing. It just sits there, waiting for the wind and the rain to batter and erode it. On the other hand, the ridiculous morsel of matter called a mouse is a wonder of the universe. It lives, it runs, it eats and reproduces itself. If ever a mighty mountain managed to bring forth a tiny mouse, we would have to declare it, not a disappointment, but the most amazing of miracles. In the words of Walt Whitman, "miracle enough to stagger sextillions of infidels."

In broad terms, the history of the universe is the story of the mountain that gave birth to a mouse. This story, chapter by chapter, is the outcome of different scientific perceptions of reality—physics, chemistry, biology, and astronomy.

That the universe had a history was a completely foreign idea to the scientists of earlier centuries. Immutable in their eyes were the laws of nature that governed the properties of matter in an eternal present. Changes such as birth, life, and death, observable in our daily lives, were explained in terms of a multitude of simple atomic reactions, themselves immutable. Matter simply had no history.

In his beautiful book on bees, the Belgian writer Maurice Maeterlinck rhapsodizes over the organization of the hive. But his enthusiasm turns

---

*"Mountains were pregnant; a ridiculous mouse was born."

to pessimism at the end when he ponders the meaning of nature and its future: "It is foolish to wonder where things and worlds are going. They are not going anywhere, and they have arrived. In a hundred billion centuries, the situation will be just the same as today, just as it was a hundred billion centuries ago, just as it has always been since a beginning which never was and always will be, until an end which is equally nonexistent. There will never be anything more, anything less, in the material or spiritual universe. . . . One could perhaps cite an experience or proof which served a useful purpose, but having traversed eternity merely to arrive at this point, does not our world demonstrate that experience is worth nothing?" The German philosopher G. W. F. Hegel expresses the same view of things in his famous saying, "Nothing new ever happens in nature."

It was with biology that the historical dimension entered the domain of science. With Charles Darwin, we discovered that animal species have not always been the same. On the surface of the globe populations changed. Humans appeared about 3 million years ago; fishes, 500 million years ago. At those moments something new did happen in nature. Life on Earth has a history.

At the beginning of this century, the historical dimension was expanded to include the whole universe with the observation of the movement of galaxies. All galaxies are moving away from each other in an expansion that embraces the entire cosmos. This fact gave birth to the idea that the universe had a beginning. Born of a fiery explosion 15 billion years ago, it has pursued its course of expansion and cooling. The image of a history of matter now reigns everywhere. Just like living creatures, stars are born, live out their lives, and die, even if their careers are measured in millions or billions of years. A galaxy in its time plays many parts: youth, maturity, and old age.

This history of the cosmos is the history of the organization of matter. The universe was born in direst penury. Nothing existed but simple, structureless particles. Like the balls on a billiard table, they were content to wander and occasionally bump into each other. Then, in successive stages, the particles combined and joined together. Their structures became more elaborate. Matter became complex and efficient.

Patience, patience
Patience dans l'azur!
Chaque atome de silence
Est la chance d'un fruit mûr!*

---

*Patience, patience/Patience in the sky-blue!/Every atom of silence/Is the chance of a ripe fruit!

The French poet Paul Valéry, stretched out on the warm sand of a lagoon, gazed at the sky. Within his field of vision, gently swaying palm trees were ripening their fruit. He was attentive to time going quietly about its business. We can apply this attentiveness to the universe. The gestation of the cosmos is reeled out on the thread of time. Every second the universe prepares something new as it proceeds slowly up the staircase of complexity.

I like to imagine a cosmic Valéry allowed to witness the unfolding of these events. It would have been his job to announce the appearance of all new entities. He would have cheered at the birth of the first atoms. In honor of the first cells, he would have composed an ode. His face would have shown signs of anxiety from time to time, for the grand cosmic ascent was not without moments of crisis. Some were serious. A few times, everything seemed lost, but, eternally inventive, the universe managed to cope. In some cases, though, it had to back up a long way before finding the right course again.

And where is this course leading? Nuclear physics allows us to understand *nuclear evolution*—how, starting with elementary particles born of the initial explosion, atomic nuclei were formed in the heart of stars. Thrown back into the immense interstellar spaces, these nuclei enveloped themselves in electrons. The remarkable progress made recently in radio astronomy and molecular biology allows us to retrace the major stages of *chemical evolution* in stars and primitive planets. And finally, following in Darwin's footsteps, we see rising up before us the great family tree of life on this planet; *biological evolution* takes us from bacteria to the appearance of human intelligence. Does the course of complexity come to a stop with the human being? We have no reason to think so.

The heartbeat of the world continues at its rhythm. Evolution follows its course. Already, perhaps on other planets, other stages have begun. What untold wonders are being prepared by cosmic gestation in each one of us? Man was born of the primate. What will be born of man?

The first section of this book is dedicated to this new idea of a history of the universe. We shall see how the observation of the cosmos has led to the view of a universe in expansion. In the light of our knowledge of the past, we shall inquire about the future of our universe. And we shall see how the most banal of observations, the darkness of the night, leads to the profoundest conclusions.

By linking nuclear, chemical, and biological evolution, we can now trace the saga of the universe to its culmination in human consciousness. In the Hindu pantheon, Shiva is responsible for the universe (figure 1). In one hand he carries fire; in the other, music. These are the two

Figure 1
Shiva, incarnation of eternal cosmic energy. A statuette from southern India,
about the 12th century A.D. In his upper right hand he holds the drum (damru),
representing music. In his upper left hand he holds fire (agni). The gestures of his
other hands reflect the eternal balance between life and death. (Rijksmuseum,
Amsterdam)

Cosmic Evolution

Nuclear evolution: from particles to atoms
In the initial furnace
In the hearts of stars

Chemical evolution: from atoms to molecules
In interstellar space
In the primitive terrestrial ocean

Biological evolution: from molecules to cells, plants, and animals
In the ocean and on the continents

Anthropological evolution

poles of the cosmos. In the beginning was the absolute rule of the flame: The universe was in limbo. Then, after countless eras, the fires slowly abated like the sea at the outgoing tide. Matter awoke and organized itself; the flame gave way to music. In the second section of the book we shall follow the stages of this birth step by step.

Warming up in the wings of evolution are some characters with familiar names: time, space, matter, force, energy, laws, chance. We should really introduce them first and define them, but we know so little about them. Every new advance in the field of physics teaches us how far we are from sounding their depths: "The best we can do is pinpoint a few islands of light in the sea of confusion" (*N1*).* In the third section we shall touch on certain questions relating to cosmic time, to forces and energy, and to the subtle connection between laws and chance. We shall catch a glimpse of a character as elusive as it is indispensable, the "elsewhere" resulting from the expansion of the universe. Without it we would not be around to talk about anything. We shall conclude with three enigmatic facts that seem to throw a surprising light on the underlying nature of matter.

As we pursue these investigations, we shall become sensitive to the underlying kinship that binds everything that exists in the universe. Man "descended" from the primate, the primate "descended" from the cell, the cell "descended" from the molecule, the molecule "descended" from the atom, the atom "descended" from the quark. We were engendered in the initial explosion, in the heart of stars, and in the immensity of interstellar space. In the purest Hindu tradition (*N2*), we can truly say that all nature is the family of man. The family links can best be shown by means of a genealogical tree, so I have put in

---

*Notes (*N*) and appendixes (*A*) will be found collected at the back of the book.

an appendix the list of names of our prime ancestors: elementary particles, atoms, and the simple molecules of galactic space (*A3*). Beyond these first generations the families multiply at an inordinate rate, so I shall limit myself to mentioning only the most influential members.

Before concluding this introduction, a word about the method used in the book. Each part is divided into chapters, each chapter into themes. These themes deal with particular aspects of the subject of the chapter. According to the reader's previous understanding, the themes will present more or less difficulty. Some could be omitted without impairing the reader's understanding of the general idea. To help the uninitiated reader get started, I have outlined at the beginning of each part the plot into which the themes are woven.

In order to relate the history of the world, I have had to call on numerous scientific disciplines. I have tried to minimize the dryness of scientific discourse by cutting away everything that is not indispensable. Still, there are some concepts that are impossible to pare down. These I have presented within a framework designed to make them more accessible. I shall try to fill my language with helpful images even though this will occasionally seem at odds with scientific rigor.

The notes and appendixes at the end of the book will serve to restate and develop certain technical points. The reader familiar with scientific language will find supplementary information there.

I have been cautious in the matter of style, resisting the temptation to polish my phrases for "literary" effect. I choose the path of innocence. The universe dwarfs us utterly, from every point of view. This is no place for affectation. Often the best approach is the most childlike, which is of course not the same as childish. It is in this spirit that I have chosen the most simplistic human-centered approach. This is because I am convinced that we cannot avoid such an approach however much we try. We can use only the logic and the language of our own age. To those who come after us we are sure to appear naive anthropomorphs. We might as well accept it.

I

The Universe Has a History

Our story begins with an exploration of the universe and an inventory of its contents. Space is populated with stars similar to our Sun. The stars that surround us are grouped into a galaxy that we call the Milky Way.

There are billions of galaxies like our own in the universe. Galaxies gather themselves into clusters, and the clusters assemble into superclusters. This hierarchical organization is one of the basic characteristics of universal architecture. We shall encounter it again at the level of atoms, and again at the level of living organisms.

In space the superclusters appear to be the last rank of the hierarchy. They follow each other tirelessly and constitute an unbounded structure that we call the universal fluid.

It is light that allows us to observe the world. But light does not propagate instantaneously. In the case of many interesting astronomical phenomena it can take millions or even billions of years to reach us. This fact profoundly influences our view of the universe. What we see are always images of the past.

Observations show that the galaxies are all retreating from one another. The universal fluid is expanding like a raisin pudding rising in the oven. What are the dimensions of the pudding? They may very well be infinite.

This expansion has been going on for about 15 billion years, which is the age of the universe. Today we know how to measure the ages of both stars and atoms. We find that the most ancient stars and atoms date back about 15 billion years. This is all pleasantly coherent.

The expansion began with a fiery blast in which matter was subjected to extreme temperatures and pressures. Radio telescopes have detected remnants of the dazzling glow that accompanied that explosion.

Other traces of the initial blast still exist. Like a hydrogen bomb, it produced atoms of helium, which are in a sense the ashes of the fire. Events taking place during the explosion might also be responsible for the absence of antimatter in our universe.

We would certainly like to know what was there "before" the initial explosion. But to do that we would have to cross over the "wall of time zero." Many formidable difficulties stand in our way, not only at the level of physics but even at that of logic itself.

It is easier to speak of the future. It may be that the expansion will continue indefinitely. It could also be that, tens of billions of years from now, it will stop and the universe will reverse its direction, leading to a period of contraction and a final implosion. The deciding factor between these two possibilities is the amount of matter in the universe. Today we have several reasons to believe that the first possibility, infinite expansion, is correct. But we could still be forced to reverse our choice.

Even if it is in infinite expansion, the universe may not be eternal. The matter out of which our bodies are formed may be disintegrating very slowly into light. Happily, the date when everything would be simply light is very far off.

This part ends with an evocation of a question of particular importance in astronomy: Why is the night sky dark? The answer is not unrelated to the expansion of the universe.

1

The Architecture of the Universe

The Realm of Stars

Lie on the ground on a cloudless night, far from any manmade sources of light. Close your eyes. After several minutes, open them to the starry vault. You will experience vertigo. Stuck to the surface of your ship, you will feel yourself a voyager in space. Savor the rapture this brings for as long as you can.

It is here that we begin our exploration of the universe. We shall try to look with new eyes, for the simplest, most immediate facts, those to which we no longer pay attention, are often the richest in information. First of all, there is the night. Half the time there is light; the other half, darkness. This is because we live very close to one star—the Sun—and very far from the myriad others.

The Sun is similar to the thousands of stars visible at night with the naked eye, similar to the billions upon billions of stars that our telescopes reveal. But, while the Sun appears to us as a blazing disk, the other stars appear only as points of weak luminosity. It is not that they are smaller or less brilliant (some are a hundred times bigger and a hundred thousand times brighter than the Sun); it is just that they are very far away. Astronomical distances are measured in terms of the time it takes light to cross them. Light crosses the Atlantic Ocean in a hundredth of a second. It travels from the Earth to the Moon in a second, and we therefore say that the Moon is 1 light-second away. It reaches the Sun in 8 minutes, so the Sun is 8 light-minutes away. In the night sky there is not a single star less than 3 light-years, or 30 billion (3×10^{13}) kilometers, away. (See, at the end of the book, Note 3 on the measurement of distances in astronomy.) Sirius is 8 light-years from us; Vega, 22 light-years; the three stars of Orion's Belt, 1500 light-years. Such are, in general, the distances between the stars. But the diameter of our Sun is only 2 light-seconds, and that of the largest star does not exceed 20 light-minutes. The heavens are empty. There is scarcely a chance that stars will collide.

The vast spaces between the stars are very dark and very cold. The interiors of stars are very hot. Between these uninhabitable domains there exist minuscule regions of hospitable temperature. Human life could not have appeared and developed except in this privileged fringe where, in rhythm with the rotation of the Earth, day and night alternate. Almost everywhere else it is always night. If there were no day and no night, we would not be here to discuss their nonexistence. But, in fact, why is the night sky black? The stars are distant, it is true, but they are also multitudinous. Why does their number not make up for their distance? This question may appear uninteresting, but it is, in fact, one of the richest we can ask. For the moment, though, let us hold it in reserve. We shall return and ponder it at length.

Let us continue our observations. We notice that the stars are not spread uniformly over the celestial vault. They are strongly concentrated along a broad bright band. In the summer this band passes over our heads like an arch. This is the Milky Way (figure 2). The naked eye cannot distinguish the individual stars of the Milky Way, any more than it can distinguish the individual leaves on the trees in a distant forest. The whitish cloud is a dense sprinkling of stars several thousand light-years away. The ensemble of these stars, including our Sun, forms the entity that we call the galaxy: our galaxy. (The Greek word *gala* means milk.) It contains more than a hundred billion stars dispersed in a volume whose shape is a disk. The diameter of this disk is 100,000 light-years, and its thickness is 5000 light-years (figures 3 and 4). Our Sun is situated about two-thirds of the way from the axis of the disk to its edge (figure 5). When we gaze at the Milky Way, what we see is a cross section of the disk, which appears to us as a narrow arch above our heads.

All the stars of the galaxy circle about its axis. The Sun makes a complete circuit in a little more than 200 million years—1 "galactic year." One galactic year ago the Earth was populated by dinosaurs. Born 4.6 billion years ago, the Sun is some 20 galactic years old.

The Realm of Galaxies

The sky as seen by the naked eye contains both stars and extended areas of luminosity called "nebulae." On the clearest nights of summer the Andromeda Nebula can be seen near the Great Square of Pegasus (figure 6). The existence of this nebula was mentioned for the first

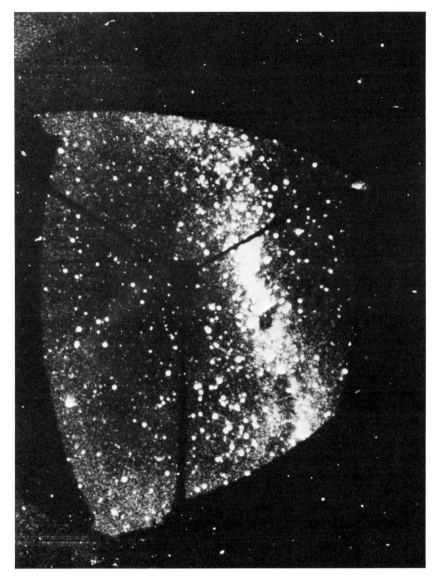

Figure 2
The Milky Way is the great luminous arch that stretches above our heads on clear summer nights. It contains the billions of stars of our galaxy, seen in cross section. This is a telescopic view.

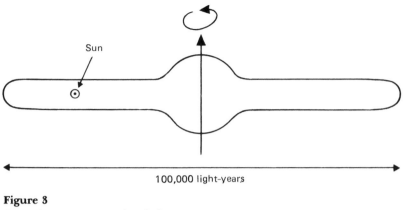

Figure 3
Our galaxy: a cross-sectional view.

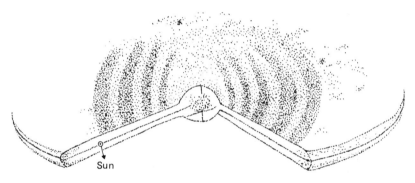

Figure 4
Our galaxy: an oblique view with a vertical cut.

time, to our knowledge, by the Arabian astronomer al-Sufi in A.D. 964. Also visible, a little below Orion's Belt, is the large Orion Nebula. But the most spectacular nebulae to the naked eye are the two Clouds of Magellan, observed by that explorer during his first voyage in the Southern Hemisphere (figure 7).

Some of these nebulae, such as the one in Orion, consist of masses of gas located within our galaxy. But the Clouds of Magellan (at a distance of some 200,000 light-years) and the Andromeda Nebula (at 2 million light-years) lie well outside the bounds of our galaxy. These nebulae are themselves galaxies of the same general nature as our Milky Way, each containing billions of stars. Immanuel Kant, in the eighteenth century, was among the first to suspect the existence of these external entities, which he called "island universes" (figures 8 and 9). But not until 1920 could the galactic hypothesis be firmly

Figure 5

Our galaxy seen face-on. This rather uncertain reconstruction of the spiral structure of the galaxy was obtained from an analysis of optical and radio-astronomical results. As everyone knows, it is much more difficult to know one's self than to know others! The approximate position of the Sun is indicated with an arrow. The absence of arms in the lower part of the figure reflects only our ignorance of the galactic structure in that region.

Figure 6
The Andromeda Nebula. This galaxy, some 2.2 million light-years away, is similar to our own. Outside the extremely bright center, in which a great number of aged (yellow and red) stars are concentrated, spiral arms develop and make many turns about the center before reaching the periphery. The internal edges of these arms consist of clouds of dust, while the outer edges are marked by chains of giant, luminous blue stars. To the right and left of the galaxy can be seen two smaller galaxies called the "satellites of Andromeda." Their amorphous structures show neither spiral arms nor young stars. We believe that these galaxies were born at the same time as Andromeda but have run the course of their lives at a very fast pace and have already attained a premature senility. (Lick Observatory)

established. With modern telescopes we can today count galaxies by the billions. The average distance between them is approximately 1 million light-years, which is not very much larger than the diameters of the galaxies (100,000 light-years). As a result, the sky is not as "empty" to galaxies as it is to stars. Collisions of galaxies are not rare.

A Hierarchical Universe

From atoms to molecules, from stars to galaxies, from clusters to super-clusters, our universe is constructed on a hierarchical plan. Similar entities join together to form new entities of higher rank. This hierarchy extends from the infinitesimally small to the infinitely large.

Prescientific human beings believed that they had been placed at the center of the universe. We now know that, at least in the geographical

Figure 7
The Greater Magellanic Cloud is an irregular galaxy. It contains a large amount
of nebular matter and many very young stars. Here the transformation of gas
into stars has not advanced far. (Lick Observatory)

Figure 8
A spiral galaxy in Coma Berenices, seen here edge-on. The disk tapers off at the edge and swells somewhat toward the center. The dark and bright pockets along the disk are clouds of interstellar matter. We think that the Milky Way is quite similar. The galaxies are retreating from each other. In 15 billion years the distances between galaxies will be twice what they are now. (Hale Observatory)

Figure 9
This elliptical galaxy has transformed practically all of its gaseous substance into stars. It shows no sign of interstellar clouds or newly formed stars. It has aged prematurely. (Kitt Peak National Observatory)

sense, they were wrong. As we explore the many schemes and structures of the universe, we shall be searching for our true place.

Since Copernicus we have known that our Earth is at the center of neither the universe nor even our own solar system. It is a quite ordinary planet revolving, like the other planets, about the Sun. And our glorious Sun is a commonplace star located in the outskirts of the Milky Way.

By observing the positions of galaxies outside our own, astronomers have established that they are not randomly distributed in the heavens. Where is the best place to look for a galaxy? Near another galaxy. And where do we have the best chance of finding a third? Near the other two! In other words, galaxies tend to form clusters, like bees or stars (figure 10). Our Milky Way is part of a cluster called the Local Group. This group consists of about twenty galaxies within a radius of about 5 million light-years. The Andromeda Nebula and the two Magellanic Clouds are also members. Here at least we are not poorly situated; our Milky Way is one of the largest galaxies in the Local Group.

Figure 10
A cluster of galaxies in the constellation Leo. (Observatoire de Haute-Provence)

Are the galactic clusters themselves organized into a higher-level structure? It certainly seems so. We call these groupings of many thousands of galaxies in volumes with dimensions measured in tens of millions of light-years "superclusters." Our Local Group belongs to the Virgo Supercluster. The central region of a supercluster is usually occupied by a monster galaxy with a mass equivalent to several hundred normal galaxies. This monster displays extraordinarily violent activity and has a number of distinctive traits. Other galaxies orbit about it in the same way that the Earth and other planets orbit the Sun, or that the Sun and other stars orbit the axis of the Milky Way. There is, however, a difference. Planetary orbits are stable, and the Earth will never fall into the Sun; but galaxies slowly approach the center of their supercluster in immense spirals. (Because they are so close and so massive, galaxies constantly perturb each other's courses.) The galaxies are irresistibly attracted by the monster cannibal, which ultimately devours them. Some astrophysicists even suspect that the monster offers these smaller galaxies as fodder to a "black hole" lodging in its midst (A6). The cannibal galaxy in our supercluster is called Messier 87 (figure 11). Is it there, on the brink of a black hole, that our galaxy will end its days? (A word of caution: These observations

Figure 11
The galaxy Messier 87 at the center of the Virgo Supercluster. (Lick Observatory)

and speculations are based on very recent studies. The criticisms of qualified experts regularly eliminate a large fraction of new theories. It is always prudent to wait a while before putting full faith in an audacious idea. Will this theory stand the test? The future will say.)

The Universe: An Endless Fluid

> At the ultimate scale the universe is a vast fluid whose elements are the galaxies, clusters, and superclusters.

The stars are grouped into galaxies, the galaxies into clusters, the clusters into superclusters. And then? Are there super superclusters made up of superclusters? Is the best place to look for a supercluster near other superclusters? It seems that this is not the case. Our result can be expressed numerically: When galaxies are more than 60 million light-years apart, they seem to ignore each other completely. But this distance is closely comparable to the size of a supercluster. There are thus, it seems, no super superclusters, that is, no groupings with distance scales larger than that of a single supercluster.

Figure 12
Each of the smudged spots in this photograph is a galaxy. They form a super-cluster located 200 million light-years from us. Such is the structure of our universe—galaxies as far as we can see. (Observatoire de Haute-Provence)

Out to the limits of the observable universe, 15 billion light-years away, the superclusters follow one another endlessly (figure 12). They are the elements of a universal fluid, just as water molecules are the elements of the oceanic fluid.

Looking "Out" Is Looking "Back"

We cannot take a "snapshot" of the universe.

Today we know that, like sound, light propagates at a fixed, well-determined speed. In 1675, while studying the motion of the satellites of Jupiter, the Danish astronomer Ole Roemer observed some bizarre phenomena, which he could explain only by postulating that light takes several tens of minutes to travel from Jupiter to the Earth. This is equivalent to a speed of about 300,000 kilometers (186,000 miles) per second, a million times the speed of sound in air. It must be emphasized that, in comparison with the distances of which we are now speaking, this speed is extremely low. On the astronomical scale, light crawls like a tortoise. The news that light carries is not always fresh!

For us, however, this is advantageous. We have found a machine that can turn back time! In looking far out, we are seeing far back in time. We see the Orion Nebula as it was when the Roman Empire fell, and the Andromeda Nebula as it was at the time of emergence of the first human beings, 2 million years ago. Conversely, hypothetical inhabitants of Andromeda, using powerful telescopes, might be watching today the origin of mankind on our planet.

The most distant objects visible in telescopes are the quasars. These are galaxies, but very special galaxies indeed. Their cores emit a fantastic amount of energy, about 10,000 times more than our entire galaxy. Each core looks from afar like a "point" source, like a star. This is the origin of the name "quasistar" or quasar. Some quasars are as far as 12 billion light-years away. The light that reaches us has been en route for 12 billion years—80 percent of the age of the universe. At the end of its incredible voyage, this light brings us scenes from the childhood of all creation.

Under these conditions, it is of course impossible to produce a "snapshot" of the universe, a picture that fixes the full landscape at a precise moment in its history. Here we sit as at the summit of the "mountain of time." In our view of the world, the most advanced point in time is the one where we are. All around, our gaze plunges into the past.

2

A Universe in Expansion

The Universe Creates Its Own Space

Our universe is spreading out, like a raisin pudding swelling in the oven, into a space that it itself creates.

Standing at the edge of a highway, with cars racing by us at high speeds, we note that the whine of each engine has a high pitch *before* the car reaches us and a much lower pitch *after* the car has passed. This is a simple reflection of the fact that in the first case the car is approaching and in the second it is moving away. Likewise, the frequency of light that an observer perceives to be emitted by an object changes according to the object's movement relative to the observer. The frequency is higher, so that the object appears relatively bluer, if it is approaching. The frequency is lower, so that the object appears redder, if it is moving away. Thanks to this phenomenon, which physicists call the Doppler effect, we can easily determine whether celestial objects are approaching us or receding from us, and we can even measure their speeds with great precision. Such measurements are now routinely carried out at a number of observatories.

The first measurements of galactic speeds were made around 1920. In 1924 the American astronomer Edwin Hubble published the astonishing results. Of the 41 galaxies observed, 36 were moving away, while only 5 were approaching. In 1929, after more cases had been studied, Hubble showed that almost all galaxies are moving away from us. The only ones that are approaching, such as Andromeda, are part of the Local Group. Beyond a certain distance, they are all receding; moreover, they are receding more quickly the farther away they are! Does this mean that we are at the center of the universe? What an appealing hypothesis, after all the dislocations humankind has suffered since Copernicus. We must, unfortunately, be disenchanted once again. To see why, let us imagine a raisin pudding sitting in the oven. The dough is rising. All the raisins are moving away from each other. Observing the movements of its colleagues, each raisin finds that, the farther away they are, the more rapidly they are receding. Each galaxy, like each raisin, has the illusion of being at the center of the universe.

Today we possess a great quantity of data on the distances and speeds of galaxies. The proportionality between distance and recession speed has been verified with high precision for speeds up to 60,000 kilometers per second (that is, up to 20 percent of the speed of light). Moreover, whether we look up, down, to the right, or to the left, we find that the expansion is proceeding at the same speed. This "isotropy" of recession speeds has also been verified to great precision. From these observations arises the concept of an expanding universe. By applying Albert Einstein's theory of general relativity, researchers have developed this concept of universal expansion from an initial explosion, or Big Bang, into a coherent theory that has now won over a large majority of scientists.

A number of observations have provided corroboration of the theory, and I shall review these in the course of this chapter. Here is just one for now. If the universe is expanding, then in the past it must have been more condensed. As we have already seen, in order to look into the past we need only look into the distance. What we observe is that the number of galaxies and quasars in a given volume *is* much greater at the most distant parts of the universe. It would be virtually impossible to explain these data if the universe were not expanding. (A specialist would quite reasonably add that one should also take into account the evolution of the galaxies themselves. Our argument remains qualitatively valid even then.)

After a lecture, a member of the audience once asked me, "Is this universal expansion real, or is it mere speculation?" It is important to understand that a whole gamut of possibilities exists between these polar extremes. A theory may be plausible, probable, very probable, almost certain, and so forth. Universal expansion, we can now say, is almost certain. (We must nevertheless present all possible interpretations for each observation, so that competing theories can be formulated and defended. In science, as elsewhere, intellectual inertia, the fashions of the moment, the weight of institutions, and authoritarianism are always to be feared. Heresies play an essential role by keeping our minds argumentative and alert. At the same time, one must maintain a certain degree of conservatism. It is unwise to throw everything into question whenever a theory encounters a small difficulty. Every scientist is to some extent a gambler before whom a number of rival theories are displayed. Weighing their strengths and weaknesses, the scientist chooses one to bet on. Unlike a racetrack gambler's bet, though, the scientist's bet is never final. The scientist is always free to change horses in midcourse in light of new observations or new calculations.)

Is the Universe Infinite?

> An impenetrable horizon removes all hope of certainty on this subject. Nonetheless, we tentatively suggest that the answer is yes.

It is not easy for the intuition to grasp the expansion of the universe. A balloon being blown up expands into the space around it. But if the universe contains all that exists, into what can it expand? Even if intuition doesn't work, though, we can still rely on our logical faculty. We have all the mathematical tools necessary for the study of an infinite fluid, whether it is expanding or contracting. Indeed, the existence of a boundary or limit would be something of a complication for our mathematics.

We are in a similar situation with respect to the geometry of four dimensions, which presents an insoluble problem to our intuition. To illustrate the three traditional dimensions, we need only extend our thumb, index finger, and middle finger in three directions. Where would we point a fourth finger? Yet geometrical problems are as easy to solve *mathematically* in four, five, or sixty-four dimensions as in one, two, or three.

Does the idea of a boundary really restore our intuition? The Greeks already showed an interest in this question. Someone suggested that the universe is surrounded by a wall. Someone else asked, "And if I climb up the wall and shoot an arrow, where will it go?"

We can try to use astronomical observations to gauge the size of the universe. But we find that, by an unfortunate combination of circumstances, these observations tell us practically nothing. There is quite simply a "horizon" beyond which we cannot see. We have already noted that the most distant galaxies are receding extremely rapidly. Certain quasars that are 12 billion light-years away are retreating from us at 80 percent of the speed of light. With more powerful telescopes one could see objects streaming outward at 90, 95, even 99 percent of the speed of light. But a beam of light emitted by a source retreating this rapidly would lose virtually all of its energy. It would wear itself out like a runner going one way on a rug that is being pulled in the opposite direction. We cannot extract information or make images from this light. And therefore we cannot "see" beyond a certain distance. An improvement in telescopes will not change the situation, for what we have encountered is not a technological problem but a fundmental issue of physics. We may thus speak of a universal or cosmological horizon. It is about 15 billion light-years away.

Let us imagine that Robinson Crusoe, knocked unconscious during his shipwreck, arrives upon his island with total amnesia. The sea extends to the horizon in every direction. The curvature of the Earth prevents him from seeing farther. He asks himself about the extent of the ocean. How far does it reach? What lies beyond the edge he can see? He might at first hypothesize that the ocean stops at the horizon. But this answer, he eventually realizes, has problems. It assumes that he is located at the exact center of the ocean, and also that what he does not see does not exist. This is a doubly egocentric attitude. After some reflection, he comes to the conclusion that the ocean *probably* extends beyond the area he can see. But then, might it not be infinite? Lacking tools to extend his knowledge, Robinson will remain in doubt.

We are going to adopt a similar attitude. Our observations are not incompatible with an infinite universe. Our intuition may get lost at such a scale, but at the local level it remains intact. Let us consider two random galaxies. All that we can state is that they are moving away from each other. The distance between them will double in another 15 billion years. Mathematically, the problem of an expanding infinite fluid does not pose any difficulties. (We shall come back to this question. Before we do, though, we shall need to introduce the concepts of open and closed universes.)

Objects at our scale of matter are characterized by both the space they occupy and the time in which they exist. Their volume extends from this to that point; their life, from this to that moment. But in the case of the universe we cannot say that it "occupies" space or that it "extends" in time. Like matter itself, these dimensions are included *within* the universe. It would seem more appropriate to say that the universe begets the space and time in which it exists. But let us be honest: We are here at the limit of intelligibility of reality.

The Age of the Universe

Galaxies, stars, and atoms all tell us that the universe was born about 15 billion years ago.

Today we have three different methods for measuring the age of the universe. These three methods are completely independent, yet they yield very nearly the same age.

The motion of galaxies
The first method rests on the observation of galaxies—more exactly, on recognition of the fact that their speed is proportional to their distance from us. Thus, if one galaxy is twice as distant as another, then the farther one is receding at twice the speed of the nearer.

To illustrate this situation, let us imagine ourselves in an observation tower perched atop the Arc de Triomphe in Paris. Boulevards stretch out in all directions, but we find that *all* the cars are moving away from us in a very particular way. Cars traveling at 100 meters per hour are now 100 meters from us, those going 50 meters per hour are 50 meters from us, and so forth. Pondering the situation for a while, we easily convince ourselves that, if their speeds have not changed, they were all gathered at the foot of the Arc de Triomphe exactly one hour ago. The cars that are now most distant are simply the ones that drove the fastest.

Let us apply this reasoning to the galaxies. We simply turn back the course of time to the moment when they were all superimposed. We find that this instant, which we could call "the beginning of the universe," occurred some 15 billion years ago. When we come to recount the history of the universe, this "time zero" will serve as the reference point for our cosmic clock.

This method is necessarily approximate. Its value is that it gives us what scientists call an "order of magnitude" answer. We can put this a little more precisely: There is a good chance that the real age will not be very different from the one this method provides.

The age of the oldest stars
We might also try to infer the age of the universe by measuring the ages of the oldest stars. This method assumes that the first stars were formed relatively quickly after the birth of the universe—a hypothesis that seems plausible in light of our present knowledge. But how can we measure the ages of stars? We simply take advantage of the fact that, like the rest of the world, they have energy problems. Their light comes from the burning of their stock of nuclear fuel. They start by burning hydrogen atoms, transforming them into helium. Then they burn their helium atoms, transforming them into heavier atoms. Stars last only as long as they have fuel left to burn. When they reach the bottom of their tank, their structure is greatly altered and they die.

Figure 13
The Messier 13 cluster. All of these stars were born together. They are about 15 billion years old. These are the oldest stars in our galaxy. (Lick Observatory)

Stars do not all live for the same length of time. The most massive stars are also the most brilliant and the shortest-lived. "Burning their candles at both ends," they survive less than 10 million years. On an astronomical scale their lifetime is that of a bonfire. Less massive stars live more parsimonious lives and survive longer. Our Sun may shine for 10 billion years, and smaller stars can have lifetimes in the hundreds of billions of years.

Stars, we observe, are born in groups, which we call "clusters." Each cluster at its birth contains the full gamut of star types, from the smallest to the most massive. In time the most brilliant stars die out. The age of the cluster is thus written in the masses of its stars; it is equal to the lifetime of the largest surviving star.

In our galaxy we find star clusters with ages ranging up to 14–16 billion years. The oldest (figure 13) are composed of "first-generation stars," which were apparently born at the same time as the galaxy. We thus have a method of dating galaxies. Applying this method to our neighbors, we find the same result: about 15 billion years. According to the theory of universal expansion, the galaxies appeared very "early," which here means within a billion years of the initial explosion.

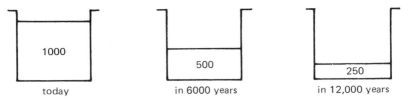

Figure 14
The half-life of carbon-14.

The ages derived from the velocity measurements of galaxies are quite compatible with this scenario.

The age of the oldest atoms
How can we measure the ages of atoms? We simply use the convenient fact that some atoms are not stable. They last for a while, and then they become transformed into other types of atoms. The best-known example is carbon-14, whose half-life is around 6000 years. The expression "half-life" needs clarification. Let us put a thousand atoms of carbon-14 in a pot and leave instructions for our descendants to count them occasionally (and to discard all non-carbon-14 atoms that they find). After 6000 years only 500 atoms will be found in the pot. After 12,000 years there will be only 250; after 18,000 years, only 125. And so on. The half-life is, then, the time required for the number of surviving atoms to decrease by half (figure 14). Archeologists have used this property of carbon-14 to date the mummies they have found entombed in the pyramids.

Today we know of more than a thousand kinds of unstable atoms. Some have half-lives measured in thousandths of a second, others in billion of years. It is to the longest-lived atoms that we now turn our attention, for with them we can date not mummies but the universe itself.

Everyone has heard of uranium. In reactors the nucleus of this atom disintegrates spectacularly, exploding into a number of particles in a process called "fission." Fission liberates a large amount of energy. The uranium is shaped into rods that are immersed in water. The liberated energy causes the water to boil, and the steam thus generated is used to drive turbines that produce electricity. In the form of a bomb this same process can devastate the countryside.

Two isotopes of uranium are of interest to us here: uranium-235, with a half-life of a billion years, and uranium-238, with a half-life of 6.5 billion years. On the Earth today there is 137 times as much uranium-238 as there is uranium-235. Since uranium-235 disappears more rapidly than uranium-238, this ratio changes over the course of the ages. When dinosaurs roamed the Earth, the ratio was 110. At the birth of the planet it was 3. Uranium nuclei, like all heavy nuclei, are formed only in the centers of stars. As the galaxy has evolved over billions of years, many uranium nuclei have been made, and many of these have disintegrated. The relative abundance of the two isotopes therefore serves as a kind of cosmic hourglass registering the passage of time.

There are many other long-lived nuclei: thorium-232 (20 billion years), rhenium-187 (50 billion years), samarium-132 (60 billion years). By measuring the abundances of these nuclei in a coherent manner we can construct a rather good chronology of past events. This is what allows us to infer from rock samples the ages of the Earth, of the Moon, and of meteorites. In all these cases we have found, with a precision of about 2 percent, an age of 4.6 billion years. This is the age we attribute, by extension, to the entire solar system. We are also able to estimate the ages of the oldest radioactive atoms. When the solar system was born, these atoms were between 5 and 12 billion years old; thus they are now between 10 and 17 billion years old.

Thus three separate methods have given us results that agree in a pleasant way. These clocks are quite different and are completely independent of each other. True, their agreement does not strictly "prove" that the universe was born at a certain time. Some people, in fact, have viewed this as a coincidence and no more. Others have offered more sophisticated explanations. Following our choice of the naive point of view, however, we cannot fail to see in this agreement a further presumption in favor of the Big Bang.

A Fossil Glow

> Light produced by the initial explosion today haunts extragalactic space. The expansion has transformed it into a feeble glow.

The most important discoveries have often been made by chance. The American astronomers Arno Penzias and Robert Wilson were using a radio telescope in an attempt to improve communications with artificial satellites. What they discovered instead was a "glow" that pervades the entire universe. Stars, galaxies, clusters, and superclusters are bathed in this radiation. It has a density of about 400 photons (particles of light) per cubic centimeter (*A1*). This light is extremely cold: its temperature is only 3 degrees absolute, which is 270 degrees centigrade below the freezing point of water.

The existence of this radiation had been predicted thirty years before Penzias and Wilson's discovery by a brilliant astrophysicist named George Gamow. Convinced of the reality of universal expansion, Gamow, in parallel with the Russian astronomer Alexander Friedmann and the Belgian priest Georges Lemaître, tried with the help of physics to retrace the course of time, like an explorer following the course of a river back to its source. They were the discoverers of the Big Bang, as Henry Schoolcraft was the discoverer of the source of the Mississippi. An indispensable guide on this journey was Albert Einstein, for in these strange realms the guideposts of Isaac Newton no longer sufficed.

Let us travel along with our explorers. In retracing the course of time, we see the galaxies approaching one another. In consequence, the mean density of the universe increases. According to the laws of physics, the temperature also increases. The past is dense and hot.

Light grows more intense as the temperature increases. Newton said that matter attracts matter. Einstein went further to assert that everything attracts everything. "Everything" includes movement; the faster a body moves, the more it is attracted and the more it attracts. "Everything" also includes bodies that have no mass, such as photons of light. Matter attracts light. Light attracts matter. Light attracts light!

Today the universe is dominated by matter, that is, by atoms, stars, and galaxies. Light carries relatively little energy. Its contribution to the universal attraction is a thousand times weaker than that of atoms. As we move backward through time, though, the situation is reversed. At the point where the universe is a billion times denser than it is today, luminous energy becomes dominant. It is here that we must abandon Newton in favor of Einstein. During the first million years of its existence, well before the birth of stars and galaxies, the universe is dominated by light.

"This primordial light still exists today," Gamow predicted. "With time, however, it weakens. The expansion has reduced it to the level of a pale glow. It should be sought not with an optical telescope, but with a radio telescope." The touchstone of a good theory is that it makes predictions, that it submits to experimental tests, and that it successfully passes these tests. Yet it was only by chance that Penzias and Wilson discovered the fossil glow thirty years after its existence had been postulated. Why was Gamow's prediction forgotten? During these years the theory of universal expansion was simply not viewed with favor. When I was a student at Cornell University, around 1960, no one ever spoke of it, and certainly no one thought of testing it. It just smelled bad. Was this because of its biblical overtones? I do not know.

In simple terms, the discovery of the fossil glow tells us that the universal expansion started from an initial state at least a billion times denser, and a thousand times hotter, than the present state. Over the ages this radiation has cooled in step with the expansion. The temperature has slowly but inexorably dropped.

From Opacity to Transparency

During the first million years of its existence the universe, because of its great density, was opaque. The light emitted in that early period was immediately reabsorbed and had no chance of making its way to us. This opacity limits our vision and eliminates all hope of "seeing" the origin of the universe (*A7*). The fossil glow was emitted at the moment of transition from opacity to transparency. It is composed of the oldest photons in existence.

What do we mean by the "transparency" of the universe? A pane of glass is transparent because light passes through it and continues its voyage in a direction outside the glass. In what direction is the universe transparent? Walking home at night, I shine my flashlight up at the sky. I send billions of white photons toward space. What is their destination? A tiny fraction will be absorbed by the air. An even smaller fraction will be intercepted by the surfaces of planets and stars. The vast majority of the photons will plod on forever. After some thousands of years they will leave our galaxy; after some millions of years they will leave our supercluster. They will wander through an ever emptier, even colder realm. *The universe is transparent in the direction of the future.*

The Ashes of the Initial Explosion

> The helium atoms that we use to inflate our balloons and the heavy hydrogen atoms are the oldest atoms that exist. They are the ashes of the original conflagration. They offer us testimony of the billion-degree temperatures that prevailed during the first few seconds of the universe.

The matter by which we are surrounded and of which we are made consists of combinations of some eighty chemical elements. We find these same elements out to the limits of the observable universe (*A3*). Numerically, hydrogen predominates: 90 percent of all atoms are hydrogen. Helium comes next with more than 9 percent. The other elements together account for only one or two parts per thousand.

The process by which the chemical elements are produced is called "nucleosynthesis" or "nuclear evolution." The goal of research on this topic is to explain, for example, why the Orion Nebula contains twice as much oxygen as carbon, and why certain galaxies are poorer in iron than ours. Where do the chemical elements come from? The majority are born in the hearts of stars; but there are some notable exceptions—in particular, hydrogen and helium. The theory of the expansion of the universe predicts higher temperatures the farther back we look into the past. Above a certain temperature level matter displays a new sort of behavior. "Nuclear reactions" begin to appear. The nuclei of atoms enter into collisions. Sometimes they combine to form new, heavier elements. Energy is emitted, generally in the form of light.

The higher the temperature, the more collisions there are, and the more violent the reactions. In the great heat of the first moments of the universe, collisions multiply without limit. Matter in its entirety detonates like a hydrogen bomb. After some minutes the temperature becomes low enough to extinguish the nuclear fire. Theory predicts that in this soup, initially composed of protons and neutrons (which, much later, become hydrogen atoms), we should find about 10 percent helium nuclei (which will become helium atoms) and 90 percent hydrogen atoms. That is not too different from what we observe today in the universe—yet another point in favor of the theory of expansion.

Was helium really born right at the beginning of the universe? Stars such as our Sun produce helium continuously. Why postulate a more distant origin? There are two reasons. First, the sum of all the helium produced by all the stars of all the galaxies is much smaller than the amount we observe. Our calculations indicate that the stars create about one atom of helium for each hundred atoms of hydrogen (instead of ten per hundred). There are some uncertainties, but it is hard to imagine that these could account for the difference. We believe today that nine out of ten helium atoms come from the Big Bang. The tenth alone was produced in the stars.

Furthermore, we observe that the heavier elements born within stars, such as oxygen, do not occur in the same abundance from one galaxy to the next, or from one region to another within a particular galaxy. That is normal enough. These abundances depend on how many stars have lived nearby—on what we might call the rhythm of a region's stellar activity. These rhythms are highly variable within the cosmos. For helium the situation is different: The proportion of eight or ten helium atoms for every hundred hydrogen atoms is *the same throughout*. This is true in active as well as in lazy galaxies, in agitated cores as well as in tranquil suburbs. The great uniformity of this result suggests a common prior cause, unique and of cosmic scale.

The genesis of nuclei during the first moments of the universal expansion explains very simply the consistency of the abundance of helium. Of course, simplicity is not in itself an ultimate criterion of truth. Things are not necessarily simple; and in many cases we know for certain that things are complicated. It is useful, though, to recall the rationale for the razor of William of Occam (a fourteenth-century theologian): If two theories explain a result equally well, it makes sense to shave away complications by using the simpler one. Other theories have been formulated to account for the abundance of helium and the uniformity of its distribution; but the number of assumptions they must make in order to achieve their results renders them relatively unattractive.

Hydrogen and helium exist in two stable forms or "isotopes." Hydrogen-1, whose nucleus is composed of a single proton, is the more common form. Hydrogen-2, called "heavy hydrogen" or "deuterium," has a nucleus composed of one proton and one neutron. It is found in heavy water and is 100,000 times less abundant than light hydrogen. For helium, the more common isotope is helium-4, with two protons and two neutrons. Helium-3, with two protons and one neutron, is about 6000 times rarer than helium-4. The initial phase of cosmic nuclear reactions generates both deuterium (heavy hydrogen) and helium-3 (light helium), as well as an isotope of lithium. The abundances calculated for these species are in very good agreement with the abundances observed in the cosmos.

To sum up, the stellar crucibles cannot account for the abundance of helium that we find everywhere we look. In order to explain it, we must postulate another hot phase in the history of the universe, since the transmutation of elements requires great heat. The movements of galaxies and the fossil glow show us the way. We must search earlier and ever earlier for this heat; it is the very source of heat we must trace. In our reconstructed scenario of universal expansion, temperatures higher than a billion degrees dominated the first few *seconds* of the universe. These temperatures drove nuclear reactions that produced between eight and ten helium atoms per hundred atoms of hydrogen, in good accord with observation. Looking at the problem from another angle, we might say that this agreement permits us to conclude that the universe was once heated to over a billion degrees.

Two Threads To Trace: The Population of Photons and the Absence of Antimatter

> According to current theories, these two threads will help us climb back even farther into the past.

In our quest for origins, we note two observational facts that are potentially rich with information. But we are not certain how to interpret them, because the theory is still incomplete. First is the fact that there are in the cosmos, on the average, a billion photons of light for every atom. Why this number rather than another? It results in large part from events that occurred in the first microseconds (millionths of a second) of the universe. No one today can say exactly what those events were.

The second thread relates to "antimatter." There are two varieties of matter: the ordinary matter of which we are formed and antimatter. Despite its somewhat dramatic name, antimatter is not fantastic; it is in some sense the twin of ordinary matter. We can imagine antiworlds, inhabited by antipeople busily going about their antilives. But—and this "but" is important—matter and antimatter must never meet. For whenever they do meet, they annihilate each other and are transformed into light.

During the first few seconds of the universe, matter and antimatter coexisted in the hot primal soup. They annihilated each other continuously to become light. They were also continuously reborn from light, like the legendary Phoenix reappearing in the midst of the flame that has consumed it. (The creation of matter and antimatter out of light and their annihilation back into light are not mere speculation. These events are of daily occurrence in nuclear physics laboratories.) During the initial epoch the populations of matter and antimatter were almost equal, to one part per billion. This incredibly small difference favored ordinary matter. At later times, when the universe had started cooling, matter and antimatter were annihilated without being reconstructed. Everything disappeared except a tiny residue, which came from the infinitesimal numerical superiority of ordinary matter. It is this bit that today constitutes all of the matter we know. Without it we would not exist. (We can confirm the absence of antimatter in the solar system, among the stars of the galaxy, and in neighboring galaxies. Strictly speaking, though, we can say nothing about the possible existence of antigalaxies at a distance of many billions of light-years.) But what is the origin of this tiny difference to which we owe our existence?

Some recent discoveries in the physics of elementary particles permit us to explore these two threads simultaneously. According to this theory, protons and neutrons—the constituents of the nuclei of all atoms—are themselves composed of yet simpler entities, called "quarks." The fusion of quarks into nucleons (protons and neutrons) occurred during the first microseconds of the universe, at a time when the temperature exceeded a million million (10^{12}) degrees. These quarks were themselves associated with the disintegration of supermassive particles that had existed much earlier still. It was amid these earlier disintegrations that the universe "chose" to become matter rather than antimatter. And it is in consequence of these events that photons have become a billion times more numerous than atoms (*N4*).

And What Was There before This?

A simple question that we cannot answer. It probably has no meaning.

A child who is wide-awake to reality discovers that the world existed before he or she did. Our birthday is not the beginning of the world, and we get used to the idea of a "prehistory" prior to ourselves. From the same perspective, it is natural to ask what came before the beginning of the universe, let us say 30 billion years ago.

In the preceding pages I have endeavored to illustrate the method of inquiry of modern cosmologists. I have compared it to the excursions into the unknown undertaken by the daring explorers of the past. We have progressed slowly, staying close to the observational evidence. We have relied upon the simplest, most naive interpretations. And we have discovered a universe that is ever warmer and ever denser, the farther back we look. The observation of the fossil glow has permitted us to see back to a million years after the beginning, when the temperature of the universe was some thousands of degrees. The measurement of the abundance of helium has permitted us to take yet another step, back to a few seconds after the origin, at temperatures of many billions of degrees. The large population of photons and the absence of antimatter have shown us ways to go back even farther in time, and even higher in temperature.

Is there any hope of advancing beyond this point?

The major problem we now encounter is that heat destroys information. When a library burns, all the data it contains are lost. In the great initial conflagration, the structures that could have recorded information were dismantled. The universe became "simple." (That is, in statistical-mechanical terms, all the distributions of positions, energies, and particles have attained a state of equilibrium. They can be described by a minimum number of parameters.) This simplicity eliminates all records. Our landmarks have vanished; we find ourselves in a world without memory.

The situation is reminiscent of the absolute zero of temperature or the speed of light. These are, for all practical purposes, inaccessible ideals. The closer one gets, the more difficult it is to proceed. From this perspective, the question "What came before?" may not make any sense. There may be no way to reach out that far (*A8*).

The Measurement of Time

It is traditional to divide time into equal periods. Then, to measure the passage of time, we count periods. This is the method of clocks. A pendulum, for example, is a clock. It swings from left to right and we count "one," and then from right to left and we count "two." The Earth is also a clock. We count "one" each time it makes a complete revolution around the Sun.

This is not the only way to measure time. We can also count "one" each time the distance between two galaxies doubles. In technical terms, we are then using a logarithmic scale (as opposed to the traditional "linear" scale of the pendulum). On this new scale, the time "zero" will be the present moment. The time "one" will arrive in 15 billion years, when any two galaxies will be twice as far apart as they are today. The time "two" will be 30 billion years later (that is, 45 billion years from now), when the galaxies will be twice as far apart as they will have been at time "one."

The past will be assigned negative times. At the time "minus one," 7.5 billion years ago, the galaxies were half as far apart as they are now. We see the most distant quasars at the time "minus four," when the galaxies were sixteen times closer together (that would be about 12 billion years ago on the traditional linear scale).

The two scales are equally useful. It is only a matter of convention, and we are free to choose whichever is most convenient. In cosmology, the logarithmic scale has two advantages. The first is of a physical nature; the second, psychological. During the early history of the universe, everything is vaporized. There is neither Earth nor pendulum to measure time. Because of the great heat, everything is accelerated. The number of reactions between particles increases for each microsecond we move back in time; events multiply toward infinity. In this sense we could say that time "slows down." The logarithmic scale expresses this situation well. As we proceed into the past, we move toward "minus infinity," which we can never reach.

The galaxies appeared at the time "minus ten" or so. To describe the period before their birth, we must redefine our logarithmic scale. Instead of speaking of the average distance between galaxies, we shall speak of the average distance between atomic particles such as nuclei or electrons. The fossil glow was emitted at time "minus one thousand." Helium appeared at "minus one billion." Quarks fused into nucleons at "minus 10^{12}." And the disintegration of the particles that gave birth to the quarks occurred at "minus 10^{27}." The psychological advantage is that, when thought of this way, there is no "beginning" of time, and therefore one is not tempted to ask what went on "before."

To the Limits of Language and Logic

"Our ideas are only intellectual instruments that help us to penetrate phenomena. It is necessary to change them when they have fulfilled their role—as one changes a bistouri when it has served long enough."
Claude Bernard

Many people hesitate to adopt the idea of universal expansion because of the philosophical and logical difficulties it would pose. The French sociologist Edgar Morin writes, "One cannot imagine a beginning out of nothing. Scientists ought to reflect on the logical problem they pose when they debate this question" (*N5*). The preceding pages provide a partial answer to this question. But there is more that can be said.

It is not only our ignorance of physics that prevents us from going back to our origins. There are also limits imposed by language. These are the limits of the scientific method and of logic itself, insofar as they must rely on language. Our words are built on the objects of our experience. They have acquired their effectiveness by adapting themselves to the occurrences of our everyday world. But when we approach realities of another scale, these words can become obstacles. Cosmology is particularly badly off in this respect, especially when it touches on questions of "finiteness" and on the limits of the universe in space and in time.

The only valid method of exploration is the empirical method. When it conflicts with philosophy or logic, it is my opinion that they are the ones that must adapt themselves to the new facts. Philosophical "difficulties" disappear if one recognizes that the only true "problem" is that of the existence of the universe itself: Why is there something rather than nothing? We cannot answer this question on the scientific level. After many millennia we are still at the same point as our ancestors, the prehistoric hunters—at absolute zero.

Our ignorance, once recognized, is the true point of departure for cosmology. There *is* something. There *is* a reality. How did it appear? What is its age? These are questions that do fall within the scope of scientific research. The problem of the existence of reality has an additional dimension: consciousness. It is by means of our consciousness that we perceive the existence of "something rather than nothing." But this consciousness is not outside of the universe; it is part of it. Today we are beginning to perceive the richness of the connections between consciousness and observational data. This discussion, however, would lead us too far from our subject.

3

The Future

The Fate of the Universe

Will the expansion continue indefinitely?

I have tried to show why we cannot easily dismiss the idea of the expansion of the universe. I am going to recapitulate the arguments briefly. The general observation that all galaxies are moving away from us, the fact that their speeds are proportional to their distances from us, and the fact that this proportionality is the same in all directions are the most obvious indications of homogeneous expansion. The other arguments are conjectural consequences of the expansion model: the increase in the density of galaxies as a function of distance; the agreement in the age of the universe as measured on the basis of the recession of the galaxies, the oldest stars, and the oldest atoms; the presence of the fossil glow; and finally the abundances of helium and other light nuclei (as well as the uniformity of the distribution of helium on the cosmic scale).

The theory has encountered certain difficulties. There are some galaxies whose abnormal motions are difficult to reconcile with the motion of the whole. In September 1976, during an international symposium at the Institut d'Astrophysique in Paris, specialists tried to come to grips with this problem. Their final report shows that abnormal cases are rare (*N6*). It is not necessary, in my opinion and in the opinion of the great majority of my colleagues, to call the theory of universal expansion into serious question. In science, it is important to remember, we do not speak of "absolute truth" or of a "perfect theory"; the role of the scientist is to gauge the relative merits of rival theories. And today the simple fact is that no other theory truly rivals that of expansion. As always, though, we must be on our guard. With new observations, the situation could rapidly evolve in unpredictable ways.

Let us now look to the future. For how long will the expansion continue? Will it stop one day, only to transform itself into a universal contraction?

The Escape Velocity of the Universe

There is a force that opposes expansion. It is gravity. Matter attracts matter, and so the universe tries to fall back on itself. This attraction plays a fundamental role in its behavior and in what it will become. To illustrate the situation, let me tell a story in the style of Jonathan Swift. On a certain planet, engineers have installed a primitive interplanetary launch pad. It is made out of an immense elastic cloth stretched over a deep valley. To launch a capsule, they place it at the center of the cloth and aim it toward their Sun. They pull the cloth down and then release it suddenly, like the string of a bow, and the capsule ascends into space. A multitude of hardworking engineers busy themselves around the base, making various kinds of measurements. They ask themselves if the launch was successful. Has the capsule risen fast enough to escape the attraction of the planet? If so, it will slip out into space and never come back. If not, it will slow down bit by bit, stop, and begin a descent that will bring it back to the launching cloth. Once more accelerated, it will rise only to descend again, like a child jumping on a trampoline. How can they tell whether the capsule has escaped or not? In ballistics, one speaks of an "escape velocity"—a minimum speed that must be imparted to an object if it is to escape the body from which it has been launched. On the Earth that speed is 11 kilometers per second; on the Moon it is 2 kilometers per second. The necessary speed depends on the force of gravity at the surface of the planet. The first thing the engineers must do, then, before they can decide whether or not the capsule will return, is to determine the strength of gravity.

In our fable, the capsule represents a galaxy, while the gravity at the surface of the planet represents the attraction of the whole universe upon this galaxy. If this attraction is sufficiently strong, the galaxies will one day cease to retreat from each other. They will reverse direction and race back toward each other, in a vast movement of universal contraction. The temperature and density will increase, and the universe will retrace, in reverse order, the great stages of the Big Bang. We call this possibility the "closed universe." Like the engineers, we ask ourselves if this is going to happen. We can imagine, in the same spirit, an infinite sequence of contractions and expansions, like the breathing of a great beast. If, however, the attraction is not sufficiently powerful to call back the galaxies, then expansion will continue indefinitely (an "open universe").

An Interminable Exhalation

Our universe seems to have too little mass for it to contract in the future.

What is our future? A new conflagration, or perhaps expansion into increasing cold, increasing emptiness. To answer this question we must calculate the gravity of the universe or, equivalently, the density of the matter it contains. "Density" is the quantity of matter in a defined volume. For example, a cube of water 1 centimeter on a side has a mass of 1 gram; the same volume of air has a mass of about 1 milligram. How can we measure the density of the universe? We must start by choosing a unit volume. It must clearly be large! Even larger than a supercluster if we are to have a representative sampling. We then take an inventory of the matter that we find in this volume. We count the galaxies, estimate their masses, and add them all up. But that is not the end. We can easily enough count the matter that we see because it sends us light, as in the case of stars. But what of the matter that we cannot see: dead stars and dead planets, asteroids far from any source of light? And what of possible forms of matter that we do not yet know or have not yet detected? How are we going to include these on our balance sheet?

Newton told us that all matter, luminous or dark, already detected or still unknown, reveals itself through gravity—through the attraction it exerts on other matter around it. By this it makes its presence known. Let us imagine, for example, that the Sun did not emit light. We would not be able to see it and would find ourselves in an eternal night. Yet the Sun would continue to attract the planets. Nothing in their motions would change. And the constellations of the zodiac would continue to pursue their annual circuit in our sky. Moreover, astronomers on Earth, without ever "seeing" the Sun, would be able to infer its existence and measure its mass by tracking the movements of the Earth relative to the stars.

By analogous methods the modern astronomer can estimate the total density of the universe, including presently "invisible" matter. We shall find it handy to express the result in terms of a volume on our everyday scale: a cube 1 meter on a side. Taking an average over all of observable space, we find that there is approximately one-third of an atom per cubic meter. This result is not very accurate. The range of possibilities extends by a factor of three on either side. In other words, the density could be as small as one-tenth of an atom per cubic meter or as large as one atom per cubic meter. There are many researchers working today to narrow the range of uncertainty of this important number.

To stop or reverse the movement of galaxies in the future, the density, according to theory, would have to be greater than *ten* atoms per cubic meter. Given the observational facts, this hardly appears possible. Because of the imprecision of our means of measurement we cannot exclude it entirely, yet the density of the universe does seem too small to bring about an eventual contraction. In that case the universe would be open.

There is another observation that seems to support the thesis of an open universe: the ratio of abundances of heavy hydrogen (deuterium) and light hydrogen. The formation of deuterium in the initial phase of nucleosynthesis depends on the density of matter. More of it is formed in a universe that is open than in one that is closed. And the ratio we observe in both interstellar space and the solar system seems closer to what we would expect of an open universe.

We have, then, two observations seeming to indicate that the expansion of the universe will continue indefinitely: the apparent "lightness" of the universe, and the abundance of deuterium. But we must be careful. The degree of certainty remains low. We lack one number that could reverse our conclusions: the mass of the particles called "neutrinos" (*A2*). Experiments to measure this number are under way (*N31*).

We must add one comment, in closing, on the dimensions of the universe. They are not the same for an open and for a closed universe. We have already noted that the open universe corresponds to the case of low density. Expansion would then proceed indefinitely, and *space would be infinite.* If the density were high, however, the universe would be closed. A definite period of expansion would be followed by one of contraction. In this case space would not be infinite. A traveler could go right around it and come back to his starting point, just like a navigator guiding a ship around the Earth. Space would then be unbounded in the sense that the celestial traveler would never encounter a frontier or a wall (any more than would the terrestrial navigator).

The idea that the universe has a history goes back to Einstein. His work on the structure of the cosmos led to the view of a universe in motion (which might be either contraction or expansion). This view drew no adherents until Hubble observed the recession of the galaxies in 1929. Since then many cosmological theories have attempted to restore to the universe its stationary, eternal character. When I was a student in the United States from 1955 to 1960, the "continuous creation" theory of Gold, Hoyle, and Bondi was generally taught. Today observation has clearly ruled that theory out. Yet the idea of an eternal universe remains attractive to many people because, they say, it eliminates the problem of the origin of matter. In fact it evades the problem, in my opinion, by sweeping it under the rug. The real problem, once again, is, Why is there something rather than nothing? Before this problem we all, scientists or otherwise, become mute. From there on all models of the universe have the right to be considered. Observation will separate the good from the inadequate. Today the weight of observation favors the view of the historical, changing universe. But the universe is what it is. It will have nothing to do with our presumptions.

The Ultimate Disintegration

Two thousand five hundred years ago the Buddha correctly anticipated the inevitable.

Today we have reason to believe that matter itself may not be eternal. More precisely we believe that the atoms of which everything is formed—shoes and ships, cabbages and kings—may one day disintegrate. How? Atoms are made up of nucleons (protons and neutrons), which are themselves made up of quarks. But the quarks themselves may not be stable (*A3*). They may eventually be transformed into radiation (*N32*). Let me reassure you, though. Ordinary atoms may not be stable, but they certainly do last a long time. They are estimated to have a half-life of 10^{32} (100,000,000,000,000,000,000,000,000,000,000) years. At that rate the Earth loses about 1 gram of matter every 20,000 years, while the Sun loses almost 20 grams a year. But time passes, and there will come a moment, in many times 10^{32} years, when no atoms will remain, no solid structures will be left. The products of this disintegration will conclude by annihilating themselves. They will become electrons, light, and neutrinos (*A2*).

According to the Hindu tradition, the universe undergoes an inevitable, periodic destruction after an interval called a kalpa. The Buddha described this interval in the following story: "Every hundred years an old man comes along to polish, with a handkerchief of the finest Benares silk, a mountain that is higher and more solid than the

Himalayas. After one kalpa the mountain will be worn down to the level of the sea." I amused myself one day by doing this calculation (*N7*). The time required is entirely compatible with the 10^{32} years mentioned above (with some uncertainties, of course). I thought that this would make a pretty and appropriate story to retell in this book.

4

Why Is the Night Dark?

The expansion of the universe is inscribed in the darkness of the nighttime sky.

The most common events are often the most mysterious and also the richest in information. "If the stars are suns, why is it that the sum of all their light does not exceed the brightness of the Sun?" This question was asked by the astronomer Johannes Kepler at the beginning of the seventeenth century, the time when the immensity of the heavens was just being realized. How far out do they extend? Are the stars uniformly distributed in space all the way to infinity? In that case they should form a dazzling screen above our heads (*N8*). *Why, then, is the night dark?* We might be tempted to respond that the stars are not in fact uniformly distributed in space. They are grouped into galaxies. This explanation is not adequate, however, because we could apply the same reasoning to galaxies, clusters, or superclusters. In the universal fluid composed of a succession of these units, the problem remains just as it was.

The correct answer to this question contains two elements with which we should now be familiar. The first is that the universe is not eternal. The second is that today the universe is transparent in the direction of the future. Clearly these two facts lead us right back to universal expansion.

All this will be easier to understand if I restate Kepler's question in different terms. Stars emit light. This energy spreads out into space like water pouring into a bathtub. Why doesn't the bathtub overflow? The idea that the universe might have an age was quite unknown to Kepler. It appeared later on with the theory of expansion, at the beginning of the twentieth century. In the context of our discussion, though, this age is not very great. Even if the heavens remained fixed, the stars are not bright enough to increase the amount of light in the nighttime sky to any appreciable extent in 15 billion years. Moreover, the heavens are not fixed but are in a state of expansion. Starlight expands into an ever vaster space. The photons emitted by the stars have practically no chance of being captured in the future. How does one fill up a bathtub whose volume is continually expanding?

In technical terms, we say that in the universe today the lifetime of photons against absorption by matter (stars or nebulae) is much longer than the age of the universe (*N9*). During the first million years of the universe, before the emission of the fossil glow, the opposite situation obtained. At that time all photons emitted were immediately reabsorbed. Kepler was right (if we replace the word "suns" with the word "particles"): The original sky *was* dazzling. It is the expansion of the universe that has caused us to pass from that period of a brilliant sky to the present dark period. It is for precisely this reason that the expansion can be held responsible for the existence of night.

The darkness of the nighttime sky has told us about the expansion of the universe. In the chapters that follow it will tell us about many other things.

To sum up, the thesis of universal expansion is in a very good position—perhaps even too good. It has almost acquired the status of dogma. The discoverers of the fossil glow have been awarded the Nobel Prize. We should distrust the aura of social acceptability accorded by such a prize. Vigilance and openness of spirit must be maintained. What troubles me personally about the Big Bang may be that it seems too simple. Can we really imagine that our world—today so extraordinarily complex and varied—was born in such a state of penury? In the next part we shall look at how the simple can give birth to the complex. But shouldn't the simple have already potentially encompassed the complex? Where were the seeds of complexity during the first minutes of the universe?

II

Nature in Gestation

"*This warm earth is perpetually in labor, possessed by the fever of gestation.*"
Fereira de Castro, *Virgin Forest*

We are going to attend a performance. Before our eyes the game of matter organizing itself will be played out. Nature, in a state of perpetual gestation, will give birth to life.

We can distinguish four major phases to this birthing process. These phases correspond to the places where the gestation occurs: the explosive universe as a whole, then the blazing hearts of the stars, then frigid space between the heavenly bodies, and finally the tepid primitive ocean.

Two preliminary sketches will help us understand the rules of the game. In the first, we shall observe what happens when a piece of iron is heated to extreme temperatures. The successive states into which it is transformed will give us an idea of the workings of different forces of nature.

Then we shall travel to the Île d'Ouessant to watch the tide go out. The behavior of the water and the reefs offers useful analogies with the movement of heat as it subsides throughout the universe.

We can also compare this subsidence with an awakening. The excessive heat of summer or of a sauna can sometimes cause a state of torpor. This was the state of the initial universe. It could do nothing. It was in limbo. Its awakening began with the partial subsidence, or remission, of heat. Then started a period of feverish activity. Various structures were gradually built. But as the inexorable decrease in temperature continued, activity diminished and then stopped. Now it was cold that caused the torpor.

We shall see this sequence of events unfold again and again. At each step one of the forces of nature is called into play. After about 1 second, the awakening is a nuclear one. The temperature decreases to a billion degrees. Because of nuclear forces, nucleons begin to combine. The first nuclei—helium, for the most part—make their appearance. But nuclear evolution is interrupted almost immediately. It does not produce the variety of heavy nuclei that will be necessary for life.

The temperature continues to go down for a million more years before the next awakening: that of the electromagnetic force. Around 3000 degrees, electrons begin to combine with nuclei to form atoms of hydrogen and helium. The atoms of hydrogen combine to form hydrogen molecules. At this instant radiation is emitted—the fossil glow that we detect today by means of our radio telescopes.

The force of gravity awakens over the following hundreds of millions of years. Enormous quantities of matter combine and give birth to galaxies. The galaxies engender the first stars. While the universe as a whole continues to cool and to become more diffuse, the stars condense and begin to heat up again.

In the centers of stars, the rising temperature reanimates the nuclear force. The stars are the reactors in which nuclear evolution goes to completion. Stars such as the Sun transform hydrogen into helium. Red giants create fertile atoms of oxygen and carbon from helium. This evolution continues during the life of the star and gives birth to all the stable nuclei, including the heaviest.

At the end of their lives, stars explode and return their matter to interstellar space. The largest become "supernovas," dying in a single gigantic burst. The smaller ones, such as our Sun, lose their material more slowly in the form of "winds."

Propelled out of the stellar furnace into the vast cold of space, the newborn nuclei acquire electrons and form a variety of atoms. This is the beginning of chemical evolution. The atoms combine into molecules and interstellar dust. Much later this dust begins to stick together to form planets in the vicinity of nascent stars. Some of these planets possess atmospheres and oceans. There chemical evolution is accelerated, giving birth to more and more complex molecules. In this broth, evolution becomes biological and produces cells, then all living creatures.

A complete narration of this performance of nature in gestation would dwell at length on this last period. I shall content myself with giving a few suggestive glimpses.

The light of the stars shows us that nuclear evolution is still occurring in every galaxy. Likewise, the molecules we detect in space, by-products of interstellar chemical evolution, prove that this process, too, is still ongoing. We are led to ask if biological evolution might not also be occurring beyond the Earth. The other planets of the solar system seem to be quite arid. Yet amino acids are found on certain meteorites; the blueprints of life were certainly sketched on these planetesimals that have since broken apart.

There are probably millions of inhabited planets in our galaxy as well as in other galaxies, but contact remains to be established.

The future of the human species depends on the future of our hospitable planet, and that in turn depends on the future of our nurturing Sun. We *can predict that, in about 5 billion years, the Sun will vaporize everything. In a semiserious vein I shall describe three possible ways to retard this event.*

We can establish an interesting analogy between the life of stars in relation to interstellar matter and the life of plants or animals in relation to compost. Two cycles of birth, death, and rebirth occur simultaneously on the Earth and in the heavens.

At the end of this part we shall ask ourselves about the music of the universe. Was it scored in advance, or is it improvised? The second possibility seems more in accord with recent advances in modern biology. Chance plays a fundamental role, but chance trained to keep only its good hits. We must admit, though, that today the music is seriously endangered.

5

The Cosmic Phase

Spectators of the Universe

Let us adopt, for a few moments, a somewhat unusual viewpoint. We shall take a step back behind time and space in order to watch, as spectators, the "movie" of the universe. Of course, we shall let ourselves be impressed by the grandiose and the brilliant. Of course, the extravagance of the masses and the energies will leave us breathless. But there is something else that will deserve our highest attention. We shall survey, with an alert eye, the appearance of structure, the achievement of successive levels of organization by matter. We shall be there to applaud each forward step, whether spectacular or, as is more likely, quiet and stealthy. There will be times when we shall become apprehensive. Crises will threaten to wipe out all of our gains. In admiration we shall watch the universe emerge from these crises and continue on its quest. Its quest of what?

Behind the scenes, other more discreet characters are at work: time, space, matter, force, heat, energy, laws, chance, information. In part III I shall describe their intrigues, their tangled interplay. For now I must just tell you a bit about them. This I shall do by describing two symbolic events: the melting of a block of iron, and the ebbing of the tide at the Île d'Ouessant.

Iron and the Fire

> Like an elevator that links the floors of an apartment building, heat gives access to the most extreme reaches of the forces of nature.

I heat up a block of iron. It begins to glow red, then orange, then white. At several thousand degrees it passes into the liquid and then into the gaseous state. The iron has now evaporated. What has happened?

We can think of the block as a sort of gigantic "molecule" composed of billions of identical iron atoms held in place in a "crystal lattice." By heating the block we are simply increasing the agitation of the atoms within it. When their thermal energy (heat) is large enough, atoms break the bonds that connect them to one another and take off on their own. The metal melts and then vaporizes. The furnace now contains an "iron vapor" consisting of isolated atoms that are free to move about in all directions. The bonds that hold the atoms in the lattice are electromagnetic in nature and result from the mutual attraction of opposite electric charges. (Electric and magnetic forces are two manifestations of a single phenomenon called the "electromagnetic force.") When, thanks to the heat of my furnace, the atoms have been sufficiently agitated to overcome this attraction, the bonds are broken. Heat vaporizes all the elements, but at temperatures that differ according to the strength of the bonds that hold them together. Dry ice evaporates at −15 degrees centigrade, water ice at +100 degrees, and metals at many hundreds of degrees (under ordinary atmospheric pressure).

Let us raise the temperature further. The heat is transmitted to the atoms of the gas and increases their speed. Their disorderly movements cause frequent collisions, and in these collisions they emit an abundance of photons. They are bathed, then, in an intense glow that they themselves produce (*A1*). Each iron atom contains a nucleus, about which 26 electrons circulate. Unusually violent collisions can dislodge electrons from their orbits, and these wander off by themselves. Occasionally they combine with other atoms for a time, then once again break away. As the temperature increases, so does the number of free electrons, until the gas is populated by completely stripped iron nuclei (bearing positive charges) immersed in a sea of free electrons (with negative charges). Such a material is called a "plasma."

We have now reached the million degree mark. The thermal agitation is terrific. More and more energetic photons are produced in the many severe collisions. Depending on their energy, these are known as x rays (as in a hospital) or gamma rays (as in a nuclear reactor). The nuclei are themselves aggregates of elementary particles called "nucleons": protons and neutrons (as in the sinister neutron bomb). These nucleons are held together by the attractive "nuclear force," which is incomparably more powerful than the electromagnetic force. Nuclei are extraordinarily stable structures. As we approach a billion degrees, though, the nuclei themselves begin to disintegrate. Attacked from all sides by energetic gamma rays, they lose protons and neutrons one by one. When the thermal energy exceeds the nuclear binding energy, the gas contains only free protons, neutrons, and electrons, all in a bath of photons.

Let us continue with our experiment, now approaching a trillion degrees. Yielding to the relentless assault of photons, the nucleons themselves begin to disintegrate. From each nucleon there emerge three quarks (*A3*). To expedite our tale, I shall refer to the force that binds quarks in nucleons as the "quark force." Strictly speaking, it is of the same nature as the nuclear force that binds the nucleons within the nucleus, although it is much more powerful. We have here reached the temperature at which the thermal energy becomes comparable to the energy of quark-force binding. At even higher temperatures other transformations will occur, but in our present ignorance we cannot specify what they will be. Experiments using particle accelerators are now trying to clear up the mysteries of quarks.

We can draw many conclusions from the spectacle we have just witnessed. First, it has shown us the activity of three great natural forces: the electromagnetic, nuclear, and quark forces. These forces have very different strengths and, as a result, do not manifest themselves under the same circumstances. Each has its own realm of activity, corresponding to a different level on the temperature scale. Next, we have discovered another hierarchy of structures. We have seen already that stars collect into galaxies, galaxies into clusters, and galactic clusters into superclusters. We now travel a similar road, but in the opposite direction. Groups are formed from components, which are themselves composed of subcomponents, and so forth. There is, however, a difference: Distances and sizes do not play the same important role here that they played in astronomy. In microscopic physics, geometrical contours are blurred. We do not speak of the "volume" of an electron.

In summary, we can describe the events that took place in the block of iron in terms of a competition between thermal energy (represented here in the chaotic motion that temperature imparts to the particles) and binding energy (due to the force by which the particles attract and hold each other). At several hundreds or thousands of degrees the electric bonds are broken, and the block of iron becomes atoms of iron vapor. At several millions of degrees the nuclear bonds are broken, and the nuclei of iron become protons and neutrons. At several trillions of degrees the nucleon binding forces are broken, and nucleons become quarks.

An Ocean of Heat

"When the tide is high, the reefs are completely submerged and deeply buried in water. They no longer exist. Only the immense sea exists, calm or mountainous at the whim of the winds.

"Then the tide begins to ebb. White areas of tormented foam and spray appear here and there. The reefs are still invisible, but the deep swells feel their presence.

"As the first rocky spurs emerge, the first breakers appear. The violent play of rock and wave, of fixed and fluid, now begins. Progressively the liquid landscape gives way to a solid landscape. The realm of the immobile—rocks, crevices, caverns—replaces the realm of the mobile—tumbling waves and cascades of white water.

"Brutally stirred up and dragged along by the currents, pebbles and grains of sand settle where they may. Precarious arrangements, constantly in flux.

"Now the reefs are almost bare. Ever shallower, ever rarer masses of water come to disturb the pebbles and sand. The movement quiets down.

"A little while, and the rocks lie drying on the abandoned beach. The landscape has wholly changed its nature. Once aquatic, it is now barren rock. And so it will remain until the next tide . . ."

I wrote these lines on the Île d'Ouessant, on the Atlantic coast of France. While watching the sea, I was struck by the analogy between the expanding universe and the ebbing tide. The sea plays the role of the initial heat. Both represent fluidity, mobility. Likewise, the architecture of the rocks symbolizes the infinitely varied structures of our present universe. When the water is high, everything is fluid, mobile; organization is abolished. At low tide the situation is reversed; the landscape is entirely mineralized. The intermediate phase is the one that matters to us. Here enough water remains for the play of gravel, sand, and pebbles. This is the period in which the landscape shows the most vitality. The analogous period plays a fundamental role in the evolution of the universe. Combinations, associations, and structures can only proliferate within particular temperature ranges. These are the fertile periods in the gestation of the cosmos. If it is too hot, everything dissociates; if it is too cold, everything crawls to a stop and mineralizes, in the broadest sense of the term.

Nuclei Emerge from the Ocean of Heat

We know little about the events that took place during the first second. Quarks seem to have played the major role. Three by three, in the first millionth of a second, they combined into nucleons. Research in this area is progressing rapidly, but for the most part we remain at the level of conjecture. The very existence of quarks has not yet been firmly established.

During the first second the universe is a great hot soup with five populations of elementary particles: protons, neutrons, electrons, photons, and neutrinos. All these particles wander about at random. Collisions are frequent and give rise to a vast range of consequences. In some cases, the collision partners simply recoil without recognizing each other. In other cases, capture occurs: A proton and a neutron combine, forming the simplest nuclear system, the deuteron (heavy hydrogen nucleus). But inevitably a photon soon appears to knock them apart again.

When the cosmic clock marks 1 second, the course of events changes. The temperature is now down to about a billion degrees. The thermal energies of the particles in the initial soup drop below the nucleon binding energies. As a result, there are fewer and fewer photons with sufficient energy to destroy the deuterons that are continually forming. The average lifetime of deuterons is extended, and their numbers increase. A new structure has made its appearance in the universe. The deuterons now set to work capturing more neutrons and protons. Nuclear systems containing three and four nucleons appear in the midst of the soup. These are the helium nuclei that will one day be used to inflate balloons. We call this period of intense nuclear activity the era of "primordial nucleosynthesis." It lasts no more than a few minutes. The temperature then drops too low to drive the nuclear reaction mechanisms. The universe congeals in its new form. It now has an abundant population of helium-4 nuclei and very tiny populations of a few other light nuclei (deuterium, helium-3, and lithium-7).

Before continuing our narrative, let us pause for a moment to reflect upon the unfolding sequence of events. We might say that in the first second the universe was in a state of rest with respect to the nuclear force, which was then inoperative, negated by the destructive effects of heat. This force was thus unable to take part in the construction of the universe. It did nothing to accomplish the nuclear plan. Thanks to the universal expansion, though, the heat is diluted, and the universe awakens to the possibilities of nuclei. Like the reefs at ebb tide, the first nuclei emerge. Some time later the landscape becomes frozen. There no longer remains enough heat for the evolution of nuclear systems to continue.

The Major Phases in the Organization of the Universe

1   **quarks → nucleons**
     (in the hot initial soup)

2   **nucleons → nuclei**
     (in the initial soup and in stellar fires)

3   **nuclei → atoms, simple molecules, and dust**
     (at stellar surfaces and in interstellar dust)

4   **simple molecules → organic molecules**
     (in the primitive ocean)

5   **organic molecules → cells**
     (in the primitive ocean)

6   **cells → plants and animals**
     (in the primitive ocean and on the continents)

We believe that analogous events occurred when, after the first millionth of a second, the first nucleons were formed from quarks. The universe then awoke from its quark-sleep. In a moment we shall witness the electromagnetic awakening. At the instant of its birth, the universe is asleep with respect to all the forces of nature. The falling temperature awakens them one by one. The universe is active for a time and then freezes; but now new structures have appeared.

The First Crisis in the Growth of Complexity

Helium does not play the game.

Some years ago astrophysicists hoped that primordial nucleosynthesis would account for the existence of all the nuclei. It was thought that successive captures of protons and neutrons might have given birth to ever more complex nuclear systems, up to uranium, in the same relative proportions that we observe today. We now know that that was not how it happened. Primordial nuclear evolution stopped at helium-4; nothing heavier was made. In a sense, the universe failed in its first experiment with nucleosynthesis. Why? Because helium is *too* stable. It was endowed at birth with extremely powerful bonds. It took advantage of these bonds to corner virtually the entire market of free neutrons. It thus blocked the progress of the game, leading to a dead end off the road to the development of complexity. At the end of primordial nucleosynthesis the universe contained only hydrogen and helium. It was sterile. Without heavy nuclei, no life could appear.

Let us pause a moment at this event, which might be called the first "growth crisis" of complexity. The associations upon which complexity depends are formed by virtue of the bonds that exist between components. The strength of these bonds is important. They should be neither too weak nor too strong. Excessive strength can cause a system to close in upon itself: It becomes "saturated," incapable of further association.

It is because helium is so strongly bound that it refuses to associate. The bonds are saturated. There are no free "hooks" to which a new partner might attach itself. The system composed of five nucleons cannot be bound; it disintegrates spontaneously. For the same reason, two nuclei of helium cannot join together to form a stable compound nucleus.

Such saturated bonds exist at many levels in nature. In chemistry we find them among the "noble gases" helium, neon, argon, krypton, and xenon. They refuse to participate in the game of chemical combination. They do not form molecules (except xenon, under certain extreme conditions). Their electrons are disposed in "complete" spherical shells about the nucleus. Once again there are no hooks to which new electrons might attach themselves. We find an analogous situation in human populations. Normally individuals group into families, families into villages, villages into countries, and so on. But if the family bonds are too strong, families close in upon themselves. No place remains for the interests of the village.

Atoms and Molecules Emerge from the Ocean of Heat

The era of nuclear combination lasts only a few minutes. The tide is now low for nuclear forces, but high for electromagnetism. Nothing happens until the thermal energy diminishes to a point where it is comparable to electromagnetic bond energies (a million times weaker than nuclear energies). This interregnum lasts a million years, the time required for the universe to cool from several billions to several thousands of degrees.

At this point the electrons and protons begin to play the same game that the neutrons and protons played during the initial phase of nucleosynthesis. A proton captures an electron, and together they form an atom of hydrogen. At the moment of capture a photon is emitted. Soon a new photon arrives to break the atom apart. But with the inexorably plunging temperatures, photons energetic enough to cause this dissociation become rarer and rarer. The atoms become less and less ephemeral, and their population continually increases. Near 3000 degrees each proton has donned an electron, and each helium nucleus is cloaked in two electrons. The universe has surmounted a new step. Atoms have been born.

And this is not all. The bond between the electron and the proton in the hydrogen atom is not saturated. Two atoms of hydrogen can join to form a "molecule," with the two electrons now circulating in complex orbits about the two adjacent protons. Thus the first molecules appear at nearly the same time as the first atoms (*N10*). Will the hydrogen molecules join with another atom of hydrogen to form a system with three atoms? Hardly. Like the helium nucleus in the nuclear plan, the hydrogen molecule is closed and rarely accepts partners. Thus we now have two closed systems characterized by the number four: the helium nucleus (two protons and two neutrons bound by nuclear forces) and the hydrogen molecule (two protons and two electrons bound by electromagnetic forces). It is not an accident that the number four is a guarantee of stability. The properties of particles occur in pairs. There are two kinds of nucleons (protons and neutrons), two electric charges (plus and minus), and so forth. When the two possible properties are both present in a system, the stability is high. When two times two properties are present, the stability is even higher. Four is a "magic" number for physicists.

The End of the Radiation Era

With the birth of atoms and molecules, other important events take place as the cosmic clock marks 1 million years. Up to this time space has abounded with free electrons that have presented a serious obstacle to the passage of light. Now there are no more free electrons. The universe has suddenly become transparent. Light can traverse it without impediment. The fossil glow that reaches us today dates from this moment deep in the past. It is made up of all these photons that, because of the transparency, will never again be absorbed (if the universe is open). (Red at the time of their birth, the expansion has degraded their energy for 15 billion years until today they have become radio photons.) At nearly the same time another event of deep significance occurs. Until now the energy associated with matter (that is to say, with the masses of particles) has been negligible compared to the energy of radiation. Now the relationship is reversed. Matter, which previously played scarcely any role in the destiny of the universe, takes over. Hereafter it is matter that will dominate the rhythm of expansion.

6

The Stellar Phase

Galaxies and Stars Emerge from the Ocean of Heat

The situation at the end of the radiation era is not ideal for the further unfolding of cosmic complexity. The systems that have already come into being, helium atoms and molecular hydrogen, are self-sufficient and refuse to play the game. Moreover, because of the expansion, the distance between atoms and molecules is constantly increasing, and they are constantly losing energy. This dispersal and cooling reduce the chance of collisions and greatly weaken the hope of further association. The advent of the reign of matter, however, will change the situation for the better. The predominance of matter over radiation will open a new chapter in the history of the universe. It is from *gravity*, not on the scale of the universe but on a variety of more local scales, that the new start will come. In the homogeneous fluid that makes up the cosmos, large masses of matter will collapse under the effects of gravitation. Here as elsewhere, there is a hierarchy. In order of decreasing size, there are the galactic superclusters, then the clusters, and then individual galaxies. We cannot be sure of the order in which these objects were formed. Some astrophysicists see progressive fragmentation: The initial fluid first divides itself into superclusters, which in turn subdivide into clusters and then finally into galaxies. Others envision the galaxies being born first, then grouping together into clusters and superclusters. Yet others picture a process starting with star clusters coming together to form galaxies. We know so little that today we must regard these and other interpretations as equally possible.

Galaxies are systems bound by the gravitational force, as helium nuclei are bound by the nuclear force and hydrogen molecules by the electromagnetic force. New condensations of matter can occur in their interiors. The distinctive nature of the gravitational force allows these gas condensations to transform part of their energy into internal heat. The tide turns, but toward *high* this time, toward increasing temperature. The gases heat up and begin to shine. We call them "stars." While the universe as a whole continues inexorably to disperse its heat, islands of matter capable of reversing this trend emerge here and there inside the galaxies. And in each of these privileged islands, the evolutionary growth of complexity resumes. After the abortive experiment of primordial nucleosynthesis, the stars are the universe's second chance.

The Life of a Galaxy

The role of galaxies is to produce stars.

How are galaxies born? By what more or less spectacular phenomena, mechanisms, and actions might a mass of matter isolate itself somewhere in the midst of the vast primal soup? How does it succeed in escaping the universal expansion, in closing in upon itself to form, in Immanuel Kant's phrase, an "island universe"? The way to find out is to look into the distance. Recall that looking out is looking back. The Space Telescope, which will be placed in orbit sometime in this decade, will permit us to take a new step back into the past. Perhaps we shall then see galaxies being born. But for the moment we are reduced to conjecture. We may picture the galactic embryos as vast shapeless nebulae, containing as much matter as several hundred billion Suns, rotating about their own centers. What is their chemical composition? They have inherited the products of primordial nucleosynthesis: hydrogen, helium, a trace of lithium, but no heavier atoms.

Inside each galaxy the force of gravity is again at work. Just as the island universes have isolated themselves from the primal soup, so the first stars are formed by separating from the original galactic material. These are the first-generation stars. They do not all have the same mass. The largest have about a hundred times the mass of the Sun and shine as bright as a hundred thousand Suns. On the galactic time scale, though, their lifetime will be very short, they burn out in less than 10 million years. Other, less massive first-generation stars will live for billions of years (*A*5). After these first stars form, others follow. Because of their mix of lifetimes, the generations overlap. As long as there is gaseous material available, stars continue to form. But after several billion years the raw material is exhausted, and the stellar birth rate falls off. Fully evolved galaxies can be recognized because they have scarcely any remaining gaseous matter and almost no young stars.

We may regard galaxies as machines for transforming gaseous matter into stars. This activity is the mainstay of their existence. But, for reasons we do not yet understand, the rhythms of their lives differ greatly. They were all born about the same time, at several hundred million years on the cosmic clock. Consequently they are all the same age today. Yet some very dynamic galaxies have already almost exhausted their gas. These are called, because of their appearance, "elliptical galaxies" (figure 9). In the lazier "irregular" galaxies the transformation has progressed hardly at all. The Magellanic Clouds (figure 7) are examples. Our Milky Way and the other spiral galaxies (figure 8) represent intermediate cases. Their rhythm is neither very fast nor very slow, and we anticipate that their stellar generations will continue to follow one another for many tens of billions of years to come.

The Lives of Stars

> Stars offer a second chance for nuclear evolution. In their cores they build up the chemical elements of life.

Galactic "compost" is much more fertile than the space between galaxies. Here and there, pulled together by its own gravity, this matter contracts and heats up. In these privileged locations we may board the thermal elevator and ascend the range of energies. We shall arrive in turn at the realm of activity of each of the forces of nature.

In the beginning, of course, it is the electromagnetic force that is activated. The first stellar embryos appear in the cores of great galactic clouds. Thanks to the motion of their accelerated electric charges, they emit the radio and infrared radiation that allows us to detect them. Under the impact of photons, their molecules dissociate into atoms and the atoms lose their orbital electrons. These electrons then wander about among the stripped nuclei.

Collisions among the particles multiply. New photons constantly appear as the energy mounts higher and higher. From the infrared, the emitted light gradually passes into the red. The star becomes "visible." According to its mass, it will eventually shade into yellow or even blue ($A\,5$).

The Fusion of Hydrogen

The thermal ascent continues until the core temperature of the star reaches about 10 million degrees. Collisions are then so violent that hydrogen nuclei (protons) can overcome the electric forces keeping them apart and can make contact with each other. We have here once again reached the nuclear stage. Nucleosynthesis, which last occurred several seconds after the birth of the universe, is now reactivated. Nucleons combine and form deuterium, then helium nuclei with three or four nucleons (*A4*). But just as before, and for the very same reasons, we can go no farther along the road of nuclear complexification. Helium refuses to play the game.

For the star, this passage to the nuclear stage represents a major step, which is manifested as a change in the star's structure. Nuclear reactions now furnish the energy needed for the emission of light. (More precisely, this energy bears the star's weight. The emission of light represents a loss of energy that menaces the star's equilibrium. To compensate for this loss, the star can either contract or burn nuclear fuel.) Now that the star no longer has to contract to keep itself in balance, it enters a novel state that we may call "stationary." Seen from the outside, it stops changing; its radius and color remain the same.

Our Sun has reached this state. The first phase of its life, during which it contracted and warmed up, lasted about 15 million years. (Astronomers speak of this as the T-Tauri, or Kelvin-Helmholtz, phase.) The Sun then entered its nuclear phase and began to transform its core hydrogen into helium. It has lived in this way for about 4.6 billion years. Throughout this time its energy flow has remained practically constant, and this constancy has certainly played a beneficial role in the development of terrestrial life.

The Sun is not alone in living through this nuclear phase. Nearly 90 percent of the stars in our night sky are similarly occupied. We may mention, among the better-known stars, Polaris, Sirius in Canis Major, Vega in Lyra, the three stars of Orion's Belt, and the four stars of the Trapezium in Orion. All the stars that are passing through this phase are said to belong to the "main sequence" (*A5*).

The nuclear phase ends when the hydrogen in the core of the star is exhausted. The Sun will reach this point in another 5 billion years. Its total lifetime on the main sequence will then have been about 10 billion years. (Remember that this lifetime is not the same for all stars. The more massive a star is, the brighter it shines, and the more rapidly it burns up its store of hydrogen. For Sirius, this phase will scarcely last longer than 100 million years. For the stars in the Trapezium, the phase will last 10 million years at most.) What happens when the core hydrogen begins to run out? Constrained by necessity, the star returns to the mode of energy production it used in its infancy. It resumes contraction. The liberated energy heats the star up again, so that it shines more and more brightly. Under the goad of gravity, the star again ascends the thermal elevator.

The Fusion of Helium, or the Miraculous Birth of Carbon

The path of nuclear evolution recovered.

At the star's heart there is now nothing to be found but helium nuclei. The temperature increases at a furious pace. It passes the hundred million degree mark. In the violence of collisions two nuclei touch for a brief moment; but nothing happens, and they separate. Helium is decidedly not sociable. Nature, however, has more than one trick up its sleeve and can now avoid repeating the failure of the first few minutes. This new trick occurs during the brief moment of contact of the helium atoms before they pull apart. Nature has arranged things so that if by chance a third helium atom presents itself at just the right moment, it can form a stable system with the other two.

This new nuclear system, formed by a sort of prestidigitation, is called carbon. The key point here is that the mass of three helium nuclei corresponds almost exactly with the mass of an excited state of the carbon nucleus. Without this agreement, apparently quite fortuitous, carbon would not have come into being. In fact, the English astronomer Fred Hoyle correctly predicted the existence and properties of this excited state based on the abundance of carbon atoms in nature.

Why did this combination not occur at the time of the initial explosion? Such a triple grouping is extraordinarily rare. The game of chance combination requires time, much time. The initial explosion lasted only a few minutes. Now, however, we can count on millions of years. And nature comes through at last! Here is the net balance sheet for the transformation: Three heliums yield one carbon (*A4*). As before, the energy given off in this reaction affects the behavior of the entire star. The contraction slows down. A new phase of nuclear fusion starts at the core, while the star's atmosphere inflates enormously and shifts toward the red. The star becomes a red giant, like Antares in the constellation Scorpio, Aldebaran in Taurus, or Betelgeuse in Orion. It is not easy to explain in simple words why the contraction of a star's core should be accompanied by an expansion of its atmosphere. It is essentially related to the variation in chemical composition between the core (helium, carbon) and the surface (hydrogen).

Over the millions of years that follow, the core of the star becomes populated with carbon nuclei. This cherished child of nature, the product of a difficult birth, is not ungrateful. Carbon will play the game of complexity on a grand scale. We shall encounter it again and again at many stages. It will become the greatest hero of chemical and biological evolution. Furthermore, in the heart of the red giant, carbon nuclei combine with helium nuclei to give birth to oxygen, another atom that will assume an important role in the organization of the world. It would not be excessive to consider the central ovens of red giants as high places of cosmic fertility.

The episodes of contraction that the star has experienced up to now have warmed not only its core but also, to a lesser degree, its exterior layers. The fusion of hydrogen propagates itself into all these regions. The star acquires an "onion-layered" structure. In the core, helium is being transformed into carbon and oxygen. Above that, hydrogen is being transformed into helium. Still higher up, nothing is changing; nuclear temperatures have not yet been achieved (*A*5). The situation is reminiscent of the variations in temperature in the great bread oven of a country baker. The pastries are judiciously placed where they will receive the ideal amount of heat: at the center the bread, then around it the tarts, and at the periphery the meringues.

The Ultimate Fusion

> Because of neutrinos, nuclear evolution takes off. A few thousand years is sufficient to give birth to nearly a hundred new chemical elements.

Helium is soon exhausted in the core of the star. Energy problems once more loom. The star contracts and begins again to ascend the thermal elevator. At around a billion degrees, it halts. Now it is carbon, the ash of the fusion of helium, that becomes combustible. Two carbon nuclei combine and give off energy. This relatively complex combustion yields many new elements: neon, sodium, magnesium, aluminum, silicon, and, in lesser quantities, phosphorous and sulfur.

At this stage an important event occurs in the life of the star. A particle, generally quite shy, makes a disruptive entrance onto the scene. This is the "neutrino" (*A*2). Shortly after reaching a billion degrees, thanks to a confluence of reactions in its incandescent center, the star starts generating ever greater numbers of neutrinos. In many ways, the neutrino resembles a photon; it has no electric charge and little or no mass. But there is an important difference. Even though the photon does not have an electric charge itself, it does have "antennas" by which it can "feel" electric charges. It belongs to the world of electromagnetism. The neutrino ignores this world. It lives in another world, that of "weak" charges. These charges are so weak, in fact, that neutrinos have only an infinitesimal interaction with the rest of the universe. The Earth, for example, is perfectly transparent to them. In the same way, the matter of a star is opaque to photons but transparent to neutrinos. While photons must laboriously clear a way for themselves from the center of the star where they are born to the surface from which they are emitted into space, neutrinos can escape the star without hindrance. In consequence, the flux of neutrinos becomes substantially larger than the flux of light. This evanescent particle dominates the life of old stars. It accelerates the emission of energy, provokes a more and more rapid contraction of the exterior layers, and prepares the way for the final catastrophe, which we shall soon witness.

After the phase of carbon fusion comes that of neon, then oxygen, then silicon. These phases occur in the temperature interval from 2 to 5 billion degrees. Because of the emission of neutrinos, they are very short. Within a few thousand years the star gives birth one after another to the nuclei of intermediate mass, from silicon to the metal group, including iron, nickel, copper, and zinc. Some of these nuclear reactions produce neutrons, which combine with the metals. Through a long chain of successive additions, we eventually see the appearance of even the most massive nuclei. Uranium-238, for example, is a nuclear system composed of 92 protons and 146 neutrons. It can take thousands of different forms, each corresponding to a particular arrangement of the orbits of protons and neutrons in its interior. It can pass from one of these "configurations" to another by emitting cascades of gamma rays. This is one of the most complex nuclear structures that exist. (Heavier nuclei would break up spontaneously under the influence of electric forces.) Nature has pursued to its limit the process of nuclear organization. Nuclear evolution, blocked in its development during the initial explosion, arrives at completion in the depths of the stellar crucibles.

The Star Explodes

In dying, the star fertilizes space with the products of its internal fires.

The stage is set when the core of the star approaches 5 billion degrees. The thermal energy now threatens to exceed the fusion energy of the nuclei. Like cakes in an oven that is too hot, the nuclear baking is in danger of "burning." The precious nuclei, patiently prepared over the lifetime of the star, are on the verge of decomposing back into nucleons. Once again disaster is imminent. The situation is saved at the last possible moment, thanks in part to the neutrinos. Their output of energy is now extremely high. To compensate for this loss, the star contracts more and more rapidly. Soon it is in free fall.

Through a series of events (which I shall not try to describe here), the contraction triggers a formidable explosion. A blazing light shines forth like a hundred million Suns. To spectators on Earth this is a "supernova" (figure 15). The layers of the onion that contain the products of the stellar furnace are hurled far away, at a speed of thousands of kilometers per second. During the months and years that follow, the stellar material, now traveling through space, retraces the evolution of the universe during its first few minutes. It becomes diluted and chilled, but with one important difference: It now contains heavy nuclei.

We may regard stellar explosions as the manifestation of a new level of cunning on the part of nature, employed to bring about increased complexity. Heavy nuclei can be born only in regions of great heat: the stellar crucibles. But we must interrupt the "baking" at the right time. We have to remove the "cookie sheet" from the oven. Such heat is completely incompatible with the formation of electric bonds. Atoms and molecules could not form at the center of a star. It is in the vast cold of space that cosmic evolution will now continue.

The Crab Nebula and the Astrologer of the Chinese Empire

After a few hundred years, the volume of expanding gas has a diameter measured in light-years. It is now a "supernova remnant." We can observe more than a hundred such remnants in our galaxy, at various stages of expansion and cooling off (figures 16–18). They are generally powerful emitters of radio waves and x rays. The Crab Nebula, in the constellation Taurus, is one of the best-known remnants. It came from a star that exploded nearly a thousand years ago. There is a pretty story attached to this event.

On the morning of July 4, 1054, the official astrologer of the Chinese empire presented himself at the imperial palace with a message of great importance. During the night a new star had appeared (figure 19). Its brilliance was awesome. Located a little above the Moon, it was as bright as Venus. On this morning, even after the Sun came up, it was still visible in the blue of the sky. The emperor received

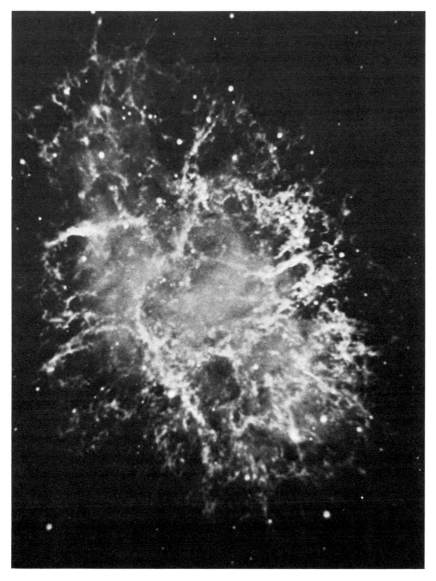

Figure 15
The Crab Nebula. Here are the remains of a star that was seen to explode on the morning of July 4, 1054. This mass of gas, previously concentrated into a star, now extends over many billions of kilometers (several light-weeks). From year to year it continues to expand in space. Its ragged contours reveal to us the violence of the movements that produced them. It is within these filaments that the heavy nuclei created in stellar fires are returned to space. (Lick Observatory)

Figure 16
Ic 443. Debris of a star that exploded several thousand years ago. (Mount Wilson and Palomar Observatories)

the astrologer and listened gravely. "What are the portents for the empire?" he asked, mindful of public welfare. "This star carries to us the promise of abundant harvests for a number of years to come," answered the astrologer. The supporting message of such a favorable horoscope was eagerly received. The newcomer was baptized the "Guest Star." Day and night it was watched. It was sketched everywhere. Parties were given in its honor. It was celebrated joyously. However, day after day its brightness diminished. For a time it could be seen only at night, like an ordinary star. Some months later it could not be seen at all. "The Guest Star is no longer here. The Guest Star has gone away," announced the astrologer. The chronicle of Chinese astronomy from which this story has been drawn does not tell us much more. Were the harvests of the following years really more abundant? Let us hope so for the sake of our astrologer. His trade was not without risk. Death was often the punishment for poorly inspired prophets.

We know today, though, that the astrologer was right. The Guest Star kept its promise. The carbon and oxygen atoms to which it gave birth will come to new harvests. But neither the emperor, nor his

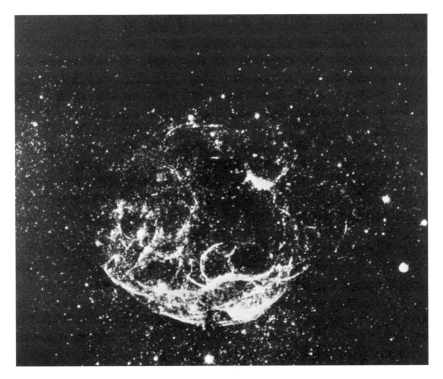

Figure 17
The Gum Nebula. A supernova remnant in an advanced state of dispersal in space. (S. Van den Bergh, Hale Observatories)

children, nor his children's children will have profited from it. Later, much later, on future planets orbiting stars yet to be born, other emperors will contemplate the fields of wheat promised by the star of July 1054, just as our harvests come to us because of Guest Stars that illumined the heavens before the birth of the Sun, and that may have been similarly received by astrologers of empires long extinguished on planets long since vaporized.

The First Catalysis

Certain go-between particles become the agents of complexity.

We shall take a few moments, before leaving this chapter of nuclear evolution, to greet the advent of the first catalysis. In the early life of the galaxy, stars are composed solely of hydrogen and helium; but in the following generations the situation changes. Thanks to the con-

Figure 18
The Cygnus Loop. A detail of filaments ejected by a stellar explosion. Here the nuclei have taken on electrons and formed atoms and molecules. Dust grains formed here may, much later, give birth to planets. This is one of the hotbeds of chemical evolution. (Lick Observatory)

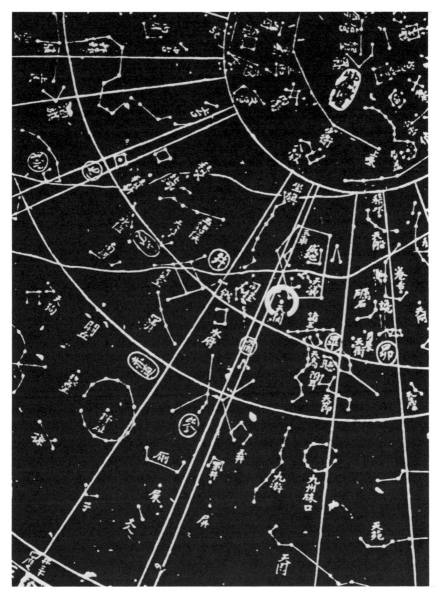

Figure 19
A Chinese map of the heavens. Meticulous in their work, the Chinese astrologers recorded all celestial events on their charts. The apparition of the "Guest Star" of July 1054 did not escape them. It is circled in white near the center of the map. Orion can be seen just below. (Miller, Lick Observatory)

tribution of the supernova remnants, interstellar material becomes progressively enriched with heavy atoms. This is an important change. Stars of later generations are formed from this enriched gas. They incorporate a certain small population of heavy atoms. Today, after 15 billion years of successive additions, this population represents less than 1 percent of the number of atoms in our galaxy. Yet these atoms play a major role. To begin with, they help in the fusion of hydrogen. In a gas containing no heavy atoms, this fusion must begin with the meeting of two protons and the formation of a deuterium nucleus. Because this reaction is always slow, fusion is relatively inefficient. In contrast, if there are some carbon atoms in the gas, fusion comes about relatively rapidly. One carbon nucleus attaches itself successively to four protons from the surrounding medium. As soon as it captures the fourth, it breaks into two parts: a nucleus of carbon and a nucleus of helium. In short, we have here the fusion of four protons into one helium nucleus, with recovery of the original carbon atom.

We are encountering for the first time "catalysis," a phenomenon that plays a fundamental role at higher levels of cosmic evolution. We can describe it as a type of matchmaking between particles. One particle plays the role of go-between. Its temporary presence has the effect of accelerating a reaction between other particles. Once this operation is completed, the go-between returns to its initial state. It can repeat this process over and over again indefinitely. Thus a very small quantity of the right catalyzing particle can play a major role in accelerating a given reaction.

In the case of the fusion of hydrogen into helium, carbon nuclei are effective catalysts only if the temperature is sufficiently high. The Sun is not hot enough, and this mechanism is not important there. In more massive (and hence hotter) stars, it becomes dominant. It has the effect of augmenting the flow of stellar energy and shortening the life of the star (*N11*). But these massive stars are precisely the ones that give birth to most of the heavy atoms of the interstellar medium. By shortening the lives of these prolific stars, carbon increases the rate of its own formation. Thus, when stars produce carbon, they alter the interstellar medium in a way that accelerates the rhythm of interstellar evolution. A second factor that has the same effect also comes into play. Heavy atoms, created in the interiors of stars, considerably enhance the opacity of the galactic gas. Matter composed exclusively of hydrogen and helium finds its normal transparency strongly diminished when even a very small quantity of new atoms is added. Opacity plays a crucial role in the formation of stars, since opaque matter condenses much more easily than transparent matter. Thus there is an increase in the rate of the transformation of gas into stars due to nucleosynthesis. *Nuclear evolution increases its own rate.*

The catalytic action of carbon has yet another beneficial effect: the birth of nitrogen. This atom, which will play an essential role in the production of the molecules of life, appears as a by-product of the catalytic fusion of hydrogen into helium (*A4*).

Stellar Residues

> "Stellar stelae" commemorate in the heavens the existence of stars that have died for the sake of evolution.

When a massive star explodes, it is not entirely dispersed into space. There is a residue. The central region falls back on itself. A new object with exotic properties appears: a star made solely of neutrons. The density of matter in a neutron star is measured in hundreds of millions of tons per cubic centimeter. This is equivalent to the mass of a large oil tanker concentrated into the volume of the head of a pin. Under these conditions, atomic nuclei crash into each other and merge. Protons are transformed into neutrons. The core of the star becomes a single gigantic "nucleus" of neutrons, held together by the force of gravity. This accounts for the name "neutron star."

The residue is also called a "pulsar" because it brightens and then fades away again many times per second. This behavior results from a combination of two effects: First, only a small fraction of the surface emits light; and second, it turns rapidly on its axis (figure 20). These are the characteristics of a lighthouse guarding a harbor. Like the lighthouse, the neutron star appears to brighten and then fade away each time we are swept by its beams of light. The first pulsar was discovered in 1964, and we have now catalogued more than a hundred. Each is a memorial to a massive star that once poured into space its cauldron full of heavy nuclei. One of the most famous pulsars was found deep in the heart of the Crab Nebula. It was formed at the time of the explosion of the Guest Star. Hardly bigger than Mont Blanc, it reminds us, thirty times per second, of the glorious event of July 4, 1054.

In certain cases a stellar residue may become even denser than a neutron star. Its gravity becomes so extraordinarily strong that it can hold its light captive and keep it from escaping. Such an object is called a "black hole" (*A6*). Do such objects really exist? We have good reason to believe that they do.

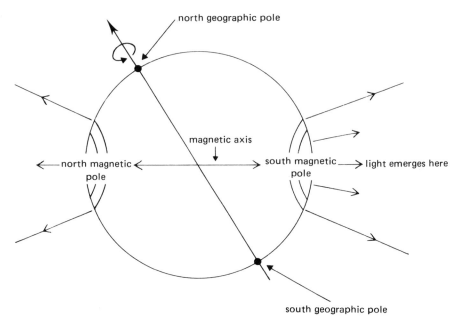

Figure 20
A pulsar. Only the magnetic poles emit light. Their rotation turns these stars into lighthouses.

The Death of a Small Star

I have described in the preceding pages the evolution and death of a massive star (figure 21). Not all stars die so dramatically. A smaller one, like the Sun, never attains the temperatures needed to bring about the explosion that eventually tears its heavier cousins apart. After its red giant phase, it also disperses the products of its internal nucleosynthesis into the far reaches of space, but in a rather mild way. It then appears in a form well characterized by astronomers as a "planetary nebula" (figure 22; *N33*). At the center of such a nebula there is usually a blue star. The nebular matter, richly tinted in yellows and reds, is the refuse of the star. Previously incorporated into the star's outer regions, it is now dissipating into space.

The denuded central star becomes a "white dwarf." White dwarfs are the residues of small stars, just as neutron stars are the residues of large stars. With respect to size, though, their roles are reversed. White dwarfs have a volume similar to that of the Earth, whereas neutron stars have the volume of a large mountain. The density of a white dwarf is in the neighborhood of a ton per cubic centimeter, in contrast to hundreds of millions of tons per cubic centimeter for neutron

Figure 21

A supernova in another galaxy. The luminous point marked with an arrow appeared suddenly in this spiral galaxy, due to the explosion of a massive star as a supernova. It does not exist in the lower photo, taken a short time earlier. Its brightness was equivalent to that of several hundred million Suns. It lasted for several months and then disappeared like the Guest Star. (Lick Observatory)

Figure 22
The planetary nebula in Lyra commemorates the death agony of a small star located close to the center of the bright ring. It has hurled out much of its own matter. The ring is made up of atoms previously located inside the star. This illustrates the return of stellar matter, enriched with new elements, to the interstellar medium. The diameter of the ring is about a trillion kilometers, that is, about one hundred times the size of the solar system. If this star possessed a family of planets, they could not have withstood the fiery blast that transported the gas. The planets would have been torn apart and vaporized. It is believed that this is the sort of fate that will overtake the Sun and the solar system in about 5 billion years (see, though, note 35). (Hale Observatory)

stars. A white dwarf has consumed all of its nuclear fuel. It cools down slowly by emitting, in the form of light, the remainder of its internal heat. Sirius, the brightest star of our evening sky, has a companion that has reached this advanced stage of stellar life (figure 23). From a white dwarf, the star slowly becomes, over the course of billions of years, a "black dwarf"—a stellar cadaver, shriveled up on itself, without radiation and without life.

The Birth of Heavy Atoms

Each nucleus, clothed in electrons, becomes an individualized atom, with a long career in prospect.

Our story has led us to the heart of the tumultuous flux that emanates from a supernova (figure 15). Heavy nuclei, created over the course

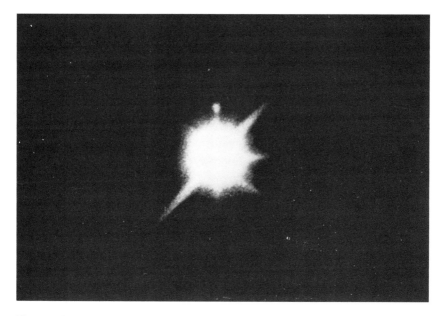

Figure 23
Sirius and its companion. A white dwarf orbits Sirius. In the picture, the light of Sirius produces the large white blob, while the companion's makes the small spot on top. The volume of the white dwarf is comparable to that of the Moon. The density of its material is immense: several tons per cubic centimeter. This is a star that has exhausted its sources of nuclear energy and is now slowly cooling down and dying. The star at the center of a planetary nebula (see figure 22) will eventually become a white dwarf. It is thus that our Sun, stripped of its planets, will end its days. (Lick Observatory)

Note: In looking at photographs of stars, it is essential to distinguish the real image from photographic artifacts. Sirius and its companion ought to form point images. The white blob and the "spikes" that point in different directions are merely photographic effects.

of a star's life, are now ejected from the intensely hot oven and hurled into the deep cold of galactic space. There they begin to capture electrons. One by one, the electrons are installed in orbits. Layer upon layer, a covering is built up around each nucleus. Each carbon nucleus acquires 6 electrons, oxygen 8, iron 26, gold 79. Heavy atoms have made their appearance in the universe (*A3*).

A sequence from Walt Disney's film *Bambi* shows us the birth of a small deer. He looks around, stretches, takes a few steps, and discovers, to his astonishment, his muscles and then his whole body. Let us imagine our newborn atoms busily discovering their structure and inventorying their potential. Some of them resemble helium. They are spherical and turned in on themselves, like a turtle or a frightened hedgehog. These are the "noble" gases: neon, argon, krypton, xenon. They will keep away from the great paths of evolution. Other atoms, however, reach out into space in complex shapes. Like stretched hands, these shapes can overlap and join with others. Each of the approximately ninety types of atoms that the universe has added to its panoply has unique properties and characteristics. These characteristics will eventually allow each atom to play specific roles assigned to it by chemistry and biology. This is true of carbon, nitrogen, and oxygen, which represent (along with hydrogen) the majority of the atoms in our bodies. It is equally true of phosphorous and sulfur, which are much less abundant but no less crucial; distributed in strategic locations, they perform indispensable tasks, for which they are irreplaceable.

A question to ponder: Does nature "know," at this moment in its history, which task it will assign to each of these newcomers? But let us return to our story, in the midst of an expanding supernova remnant. Our new atoms are now meeting their companions. They try each other out in a game of molecular liaisons. Oxygen, in particular, has a great deal of success. It forms quite stable and durable bonds with a great variety of other atoms, especially metals such as aluminum, titanium, iron, magnesium, and silicon. These are the first oxides. They will serve as a basis for all solid constructions in the universe.

The Birth of Crystals

The first solids come into being in the ashes of exploded stars.

The game of molecular combinations continues for a long time in the heart of the remnant nebula, where it eventually produces the first crystals. The phenomenon is not simple, but we can show its central features in a schematic diagram (figure 24). Let us imagine that at a

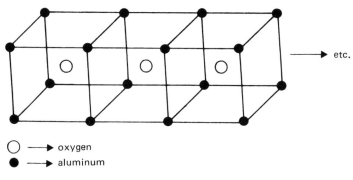

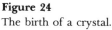 oxygen
● ——→ aluminum

Figure 24
The birth of a crystal.

given moment we have arrived at the following situation: Around an atom of oxygen are arranged eight atoms of aluminum. They occupy the corners of a cube with the oxygen atom at the center. Then, around each aluminum atom, eight oxygen atoms are dispersed in similar fashion. And then again eight aluminums. This "pattern" is repeated indefinitely, as in certain tapestries, but here in three dimensions. The structure thus created will contain a large number of atoms. It is a "crystalline lattice" or, more simply, a crystal. The solid matter that surrounds us is largely composed of crystals. The nature of the crystal is fixed by the chemical elements it contains. In a salt-shaker, the grains of salt are crystals of chlorine and sodium. Quartz is made of silicon and oxygen. Some crystals are made up of only one substance; a diamond, for example, is a lattice of carbon atoms. Our ordinary terrestrial rocks, on the other hand, are made up of complex networks of oxygen, silicon, magnesium, iron, aluminum, and other atoms.

The Secret of Purity

The growth of crystals has been much studied in the laboratory. Even if the medium in which they grow contains a large variety of different atoms, what we see are always tiny, very pure crystals made up of a combination of just two or three types of atoms, to the exclusion of all others. These crystals pursue a course of growth that preserves their purity (and hence their identity). Let us consider a quartz crystal, made up of oxygen and silicon, forming in a liquid medium. Surrounded by that medium, our crystal is continuously bombarded by all the types of atoms present. A few stick to its surface. If they are oxygen or silicon atoms, and if their placement corresponds exactly to what is necessary for the growth of the crystal, they become solidly attached. They are now part of the crystal and contribute to its further growth. If a new arrival is foreign to the crystalline lattice, however, it will not find a suitable lodging. Like a key that does not fit into a lock, its geometrical form will not adapt itself to the atomic architecture already established. It will be thrown back into the liquid. In the world of crystals, geometry is the password.

Individual atoms do not possess the selectivity shown by crystals. Selectivity appeared as a result of the juxtaposition of the first atoms. It is an example of an "emergent property" in the organization of matter. Such a property itself gives birth to a novel action—in this case, a prototype of feeding. Like a living being, the crystal can incorporate matter selectively by not retaining whatever comes to it unless the new material will help preserve its identity. In a sense, then, we could say that it feeds itself.

7

The Interstellar Phase

Interstellar Dust Grains

Ice-covered rocky particles foreshadow the planets and give them birth.

Space is populated by myriads of grains of solid matter called "interstellar dust." These grains are comparable in size to particles of smoke; they are less than a micron (one thousandth of a millimeter) in diameter. But on the atomic scale they are like mountains. Each contains trillions of atoms. This dust is formed, among other places, in expanding and rapidly cooling gaseous masses. Supernova remnants are involved, but so are the less spectacular explosions of novas. The gaseous envelopes of red giants and planetary nebulae are other possibilities. For our purposes, let us just remember that the first such dust appeared, so far as we can tell, among the torrents of gas that were hurled into space when the first-generation stars died (figure 18).

Hydrogen Enters the Game

After a few tens of thousands of years the supernova remnant has spread itself out into space. It now occupies a volume some tens of light-years in diameter. The Cygnus Nebula is a good example. Its garlands of color are wound up like the scrolls of smoke from a cigarette (figure 18). The temperature of the remnant now approaches that of the interstellar clouds, only a few tens of degrees above absolute zero. Not everything is dead out there, though. Chemical activity resumes, vigorously. It is hydrogen, this time, that leads the way. Joining with the heavy atoms, it forms several new molecules that are quite familiar to us: water (hydrogen and oxygen), ammonia (hydrogen and nitrogen), and methane and other hydrocarbons (hydrogen and carbon).

Some of these molecules contain a particularly versatile type of joint called the "hydrogen bond." It is a property specific to hydrogen that allows it to create these strong bonds, which allow molecules of liquid water to attach themselves to each other, resulting in an unusually high boiling point. Without the hydrogen bond, the oceans would have rapidly evaporated and life would not have appeared on the Earth. These hydrogen-containing molecules form a thin icy shell around the dust grains. Now, thanks to the action of the ultraviolet radiation from neighboring stars and the cosmic rays that are propagating throughout space, a new chapter in our story is going to begin: the formation of complex molecules. But before that, I must add a few words on cosmic rays.

Cosmic Rays

> Extremely rapid particles stream through space in all directions. They participate in nuclear evolution, in chemical evolution, and in biological evolution.

"Cosmic rays" were discovered at the beginning of the century by physicists studying radioactivity. "Radioactivity" is the name given by Henri Becquerel and Marie and Pierre Curie to the disintegration of unstable nuclei. During this disintegration, the nuclei emit very fast particles. Special detectors were invented to study these particles, and one day someone noticed that the detectors continued to register weak impacts even in the absence of radioactive sources. What was the source of this "background noise"? A long inquiry, worthy of Hercule Poirot at his best, revealed that the noise was coming from galactic space. Up there, between the stars, many particles are racing around at close to the speed of light. Some of them are electrons, some are protons, and some are complex nuclei. A few of these particles enter the solar system, reach the Earth, penetrate our atmosphere, and end their days in radioactivity detectors. Such particles are collectively called "cosmic rays."

What can be said about the origin of these fast-moving particles? Some were accelerated in each of a large number of violent phenomena that have occurred in the cosmos. We know (because our satellites have detected them) that some are created in the eruptions that burst out sporadically from the surface of the Sun (figure 25). Supernova explosions are another major contributor. Other even more violent events, such as the prodigious activity of galactic cores, could also play a role.

Figure 25
A solar flare. Within just a few hours, regions of the Sun's surface can become active and eject immense tongues of hot gas into space. Intense fluxes of fast particles emerge from these flares and flow away, to Earth and beyond. Here is one of the sources of cosmic radiation, which plays a very important role in cosmic evolution. (Sacramento Peak Observatory)

After being accelerated, cosmic rays wander randomly through the galaxy. They rush blindly into whatever they encounter along their way. Their collisions give rise to a number of reactions that are important for our cosmic epic, especially in the context of nuclear evolution. Cosmic rays have energies many times higher than what is necessary to break nuclear bonds. Under the violence of their impact, some interstellar atomic nuclei will be broken into pieces. These fragments are smaller nuclei that will surround themselves with electrons to form new types of atoms. Among these new nuclei, three were lacking in the previous panoply of the universe: lithium, beryllium, and boron. These are very fragile and cannot tolerate high temperatures; as a result, they do not form in stars. Cosmic rays thus complete the process of nuclear evolution by adding these three chemical elements (*A4*). In the realm of biological evolution, it is generally believed that the impact of cosmic rays on molecules of the genetic material can cause the mutations that are the driving force behind Darwinian evolution. We shall come back to this point later.

In space, on the surface of dust grains, the cosmic rays help to initiate chemical evolution. They start by breaking up the molecules that form the icy shells. The fragments then recombine randomly. This is the start of a new chemistry. Molecules unknown until this moment begin to form. Through these dissociations and conjunctions, some molecules attain significant dimensions; they can include more than ten atoms (*A3*). Undoubtedly some are even larger, but the present state of our observations and extrapolations leaves us little hope of finding true "macromolecules" in space.

The bonds that attach these molecules to the dust grains are weak. Like children now grown up, they leave the land of their birth and travel elsewhere to live out their lives.

Interstellar Molecules

Every type of molecule emits a unique electromagnetic spectrum — its "signature" — in the form of radio waves. By recording these signatures, our radio telescopes have allowed us to identify nearly a hundred different interstellar molecules (*A3*). These observations cannot be overemphasized, for they have had an important impact on astronomy as well as biology. We certainly expected to find a few simple molecules in space, but because of the extreme conditions of temperature and density, no one foresaw the possibility of such an array of complex molecules. We stand amazed before this frenzy of organization. Matter seems able to draw advantage from even the most adverse circumstances.

The ancients believed in spontaneous generation. They thought that decaying garbage could give birth to flies and even rats. Louis Pasteur finally destroyed this belief, showing conclusively that life always comes from life. But where did the first life come from? It is well accepted that life arose from "inanimate" matter at a very remote time in the Earth's history. Until recently this seemed to have been an exceedingly improbable event — in fact, a kind of miracle born of a juxtaposition of coincidences, each more extraordinary than the last. Since this "miracle" could not have occurred more than once in the universe, it was generally believed that we had to be totally alone. This is, for example, the theme of Jacques Monod in *Chance and Necessity*. Our unanticipated detection of the profusion of interstellar molecules has now cast doubt on this belief. It has blurred the distinction between "inanimate" and "animate" matter.

What is life? In the larger sense the word denotes the mysterious tendency of matter to organize itself and to exhibit levels of complexity.

Certainly the molecular activity that takes place on grains in space is as much a manifestation of this tendency as the proliferation of tropical vines in the Amazon or the nucleosynthesis that occurs in stellar furnaces.

Recall for a moment the list of molecules that populate the great interstellar clouds. Note that all of the ones that incorporate more than three atoms include one, two, or three atoms of carbon, a structure equipped with four "hooks" particularly well adapted for molecular combination. On the Earth, too, carbon is present in all large molecular structures. What we are discovering, then, is that the predominance of carbon is not confined to the biosphere. It permeates our galaxy and without doubt others as well.

Could there exist a life form very different from what we know on the Earth—a life form in which, for example, silicon plays the role of carbon? At first glance there is nothing to oppose this idea, since silicon has the same four electronic hooks as carbon. Yet the list of molecules in space includes very few that contain silicon, as opposed to many tens of molecules that contain carbon. Why? Undoubtedly this is due to the fact that the bonds that connect silicon atoms are much stronger than those involving carbon. Whatever it grabs onto, silicon cannot release again. Like helium, it is incapable of playing the game. The observation of its scarcity by our radio telescopes therefore makes the existence of planetary life based on silicon very unlikely.

8

The Planetary Phase

stand back
shoulders against the sky
wait for the Earth
Pierre Dubois

Inventing the Planet

Myriads of interstellar dust grains come together to create a land of supreme fertility.

Admittedly, conditions in interstellar space are difficult. It is cold out there, atoms are rare, and the probability of encounters and combinations is small. Moreover, the new molecules are fragile and menaced by ultraviolet light and cosmic rays, which at this stage become hostile to evolution. Molecular proliferation is obstructed by this "ionizing radiation." Lacking resistance to the bombardment, the molecules need shelter.

Nature has invented something new to overcome this difficulty. What is needed is a medium that is neither too hot (the molecules would dissociate) nor too cold (the molecules would ignore each other). The medium should be dense enough to facilitate collisions and to protect molecules from the lethal rays that abound in space. The invention is called a "planet." It must be situated near a star that can supply it with energy. Attached to the star through the force of gravity, a planet in circular orbit can stay at a distance where the temperature is moderate. Moreover, if its mass is sufficient, the planet will have a gravitational pull that can retain layers of gaseous substances, which will constitute its atmosphere and isolate it from the radiation of space.

The Birth of Planets

Our knowledge of planetary formation comes from two different sources: the astronomical observation of stars being born, and the exploration of our own solar system.

The first source of information can be useful to us only if events occurring today are similar to those of 5 billion years ago. This seem likely. When our Sun was born, the galaxy was already 10 billion years old. Thousands of generations of stars had succeeded each other in the galaxy, shaping it in form and substance. The face that the galaxy presented then was not much different, so far as we know, from the face it presents now. But the galaxy is vast and the nurseries of the stars are distant. Our instruments do not yet have the resolving power required for detailed observation. We must content ourselves with fragmentary and incomplete information. What have we learned? That stars are born in groups, deep within the great galactic clouds, amidst dust and interstellar molecules (figure 26).

The study of planets, satellites, and meteorites provides our second source of information. In this case, the astrophysicist becomes an archeologist, searching for clues from the past. We try to identify objects that have retained intact a memory of the origin of the solar system. On the Earth, everything moves, everything changes. Traces of the past are quickly erased. Inert celestial bodies, without atmospheres and without volcanic activity, are much more eloquent. Examples include both our Moon and the meteorites that fall on us from the sky.

Combining the information so far collected, we can form an approximate idea of the sequence of events. Let us return in our imagination to the heart of a vast opaque nebula, made up of gaseous and dusty matter. This dust sets itself in vast sheets around the first stellar embryos, rather like the rings around Saturn. A long process of condensation is initiated in these disks. The grains of dust begin to stick to each other. Many small bodies result, moving in more or less regular orbits, gravitationally bound to the central object. Their orbits often cross, and collisions are frequent. Depending on the violence of the impact, the small bodies either fragment or combine. There follows a period of competition, in the course of which certain objects see their masses increase. The larger bodies absorb the smaller ones. Along with their masses, their gravitational fields increase. Around the most massive bodies, the accretion proceeds like an avalanche. The system depopulates itself to the profit of the big winners, the planets. But the final avalanche has left its mark. The fall of small

Figure 26

This photo shows several stars obscured by filamentary nebulosity. These are the Pleiades. They are a small cluster of stars that are visible to the naked eye in the winter sky. They were all born together from an interstellar cloud about 100 million years ago. The filaments are composed of interstellar dust particles. They resemble the cirrus clouds of our own atmosphere, made of tiny ice crystals. This dust, here seen illuminated by the light of the stars, may collect to form planets orbiting stellar embryos. The bright radial spikes around the stars are photographic effects of no interest to us. (Lick Observatory)

Figure 27
The surface of Mercury is pocked with craters. These are the scars left by colli-
sions of the planet with smaller bodies at the time of origin of the solar system.
(NASA)

bodies onto the forming planets has pocked their surfaces with craters
that can still be seen. Under their impact, the ground melts, and
incandescent liquid rock sprays great distances. When the rock solidifies,
it leaves a crater. Small or large, the craters overlap and bury each
other. The disorderly mosaic of craters forms the background for the
surface relief of Mercury and the Moon (figures 27 and 28).

The Heat of Planets

An inheritance from the protosolar nebula, heat is the motor that drives
planetary life.

Our exploration of the solar system has helped us realize the dominant
role that heat plays in the life of a planet. But what is the origin of
this heat? There are in fact two distinct sources: First, the violent
collisions of the meteoritic avalanche, and second, the natural radioactivity of unstable atoms such as uranium and thorium from the
original nebula. These unstable atoms, created in massive stars and

Figure 28
Astronauts have found only unrelieved aridity on the Moon. It is too light to retain any gaseous substances that it may originally have possessed. (NASA)

ejected into space as supernova debris, are incorporated into dust grains and from there find their way into planetary material. They then disintegrate in their own good time, releasing their energy to the crystals in which they have become embedded.

The more massive a planet, the more heat it will inherit from these sources. At their birth, planets are balls of incandescent lava. With time their heat dissipates into space. Smaller bodies—asteroids and meteorites—cool off very quickly. After that they are frozen for eternity, and must be content to record in the form of craters the violent shocks of later collisions. The Moon, eighty times lighter than the Earth, became frozen some 300 million years after its birth. Neither crustal movement nor volcanic activity disturbs its surface today. Mercury, which is a little heavier, lived a few hundred million years longer before reaching the state of total solidification that it now displays. Mars is an intermediate case between the Moon (or Mercury) and the Earth. It has largely, but not completely, dissipated its initial thermal endowment. Its rare volcanoes testify to this fact (figures 29 and 30).

Under the formidable meteoritic avalanche that gave it birth, our planet seems to have remained in a liquid state for many millions of

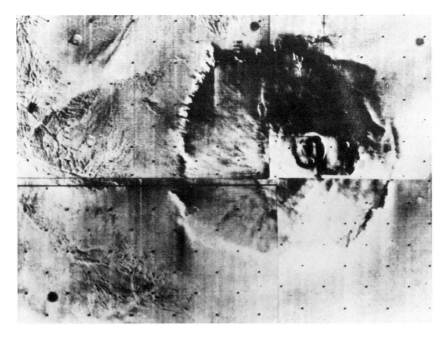

Figure 29
Martian volcanoes are rare but much larger than those on Earth. This one, Olympus Mons, rises 23 kilometers, and its base extends over an area larger than Belgium and Holland combined. These volcanoes work to dissipate the interior heat accumulated since the time of formation of the planet. (The grid of small dots is superimposed by the camera for measurement purposes.) (NASA)

years. The first stable crust did not appear until the end of this period. The fluid interior, stirred by powerful convective motions, continued to seethe for a long time. It is this heat that still animates volcanoes, earthquakes, and the continuing continental drift. The Earth is a prototype of the living planet.

The Birth of the Atmosphere

Interstellar dust provides us with our atmosphere and our oceans.

As our planet in the process of formation sweeps through space in its early orbit, it gathers up all the stones and dust that cross its path. This dust—rocky nuclei surrounded by layers of ice—is devoured by the red incandescent ball (figure 31). What becomes of these specks in the molten mass? They first evaporate and then dissolve in the interior of the Earth.

Figure 30
An automatic laboratory on the surface of Mars. These rocks, when analyzed by chemical methods, showed no trace of life, however rudimentary. Complex molecules were practically nonexistent. (NASA)

Liquid rock can incorporate large quantities of gaseous material, but solid rock cannot. When the first terrestrial crust is formed, openings appear, like volcanic rents. In mighty geysers masses of gas escape to the surface. (This is like the outgassing that occurs when carbonated water freezes. I learned this to my sorrow once when I forgot about a bottle of champagne I had left in the freezer. The effects were impressive.) The planet becomes clothed in a vast, dense atmosphere (figure 32). Water begins to condense. It rains as it will never rain again. It rains entire oceans.

Water! Water!

On the cosmic scale, water is rarer than gold.

In school I learned that matter exists in three forms: solid, liquid, and gas. I also learned that the oceans cover 70 percent of our planet. At the time of the hypothetical initial partitioning among these forms, the liquid phase seems to have been particularly favored. Seen from

Figures 31 and 32
Eruption of the volcano Surtsey in Iceland. This is what the first few million
years of the Earth's history may have looked like. Vast numbers of volcanoes released molten rock and water vapor. Ice brought by interstellar dust particles vaporized when the planet formed. It was released and then fell back onto the
surface to make oceans that completely covered the planet. The primitive ocean
was a high place in cosmic fertility. (Solarfilma, Reykjavik, Iceland)

space, the situation is different. On the scale of the galaxy or universe, nearly all matter is gaseous, either neutral or ionized. Solid matter forms no more than one-millionth, and liquid matter no more than one-billionth, of the universe. (I have made these estimates using plausible hypotheses regarding the frequency of planetary systems. I have not included the quasicrystalline structures of the white dwarfs.) Old-time whalers at sea spent two or three years without seeing anything except the unlimited expanse of the oceans. Would they have believed that on the cosmic scale liquid water is rarer than gold is on the Earth?

It is difficult to overestimate the importance of water in all its forms for cosmic evolution. Its dissolving power allows it to incorporate large quantities of other molecules, which then circulate freely in solution. The opportunity for encounters is multiplied, and contacts are prolonged. From this point of view, water is a powerful aid to organization. Let us take this opportunity to salute the appearance of liquid water on the Earth. This is certainly not the first time the event has occurred in the universe. Presumably it occurred on many other planets that vanished long before the Sun formed. But symbolically (and egocentrically), we choose to celebrate its arrival upon our Earth.

The Great Flood

From our chosen perspective as spectators, we now see great events happening in the organizational scheme of matter. At the time of the initial deluge, the Earth looked somewhat like the planet Venus does today (figure 33). From space, during the flood, we cannot see the continents, which are being slowly inundated by water; a dark, opaque, gaseous mass covers the entire surface of the planet. Transported by cyclonic motions of great power, this gas takes on the banded appearance that is found today not only on Venus but also on Jupiter and Saturn. The space probes that descended into the atmosphere of Venus in December 1978 underwent a continuous bombardment of electrical discharges, presumably as spectacular as our lightning. We have every reason to believe that our primitive atmosphere, which was as dense as that of Venus today, nurtured similar phenomena. The lightning flashes were accompanied by sound waves that reverberated endlessly. The atmosphere was filled with a thunderous rumbling, to which, fortunately, no ear had to listen. (Why doesn't Venus have any oceans? In the beginning, its atmosphere closely resembled that of the Earth. The difference seems to have been a matter of temperature. Venus is nearer to the Sun than the Earth is. Each square

Figure 33

Venus. We cannot see the surface of Venus from space. We see nothing but the motions of clouds entrained by the revolution of the atmosphere. Earth, in its earliest history, may have had a similar appearance before its water, condensed as rain, fell and collected in the ocean basins. (NASA) meter of its surface receives twice as much heat as a comparable area on Earth. Observations made by the Venus probes lead us to hypothesize that droplets forming high in the Venusian atmosphere never reach the ground. They evaporate as they fall, because the mean temperature is much higher than the Earth's. It never rains on Venus.)

The molecules in the primitive atmosphere (carbon dioxide, methane, ammonia, water vapor, and apparently several other more massive species inherited from interstellar space) are continuously bombarded by electric discharges and also by the ultraviolet radiation of the young Sun. Recall the interstellar dust grains, surrounded by ice and subjected to cosmic rays. The dissociated molecules recombined to produce some real chemistry. However, this organizational activity did not go very far under the inhospitable conditions of space. The most complicated molecules we can detect in interstellar clouds contain no more than about a dozen atoms. Yet these interstellar dust grains in some ways foreshadow planets like our own, with rocky interiors and oceans.

Photochemical evolution begins again in the primitive atmosphere, but under much more favorable conditions. First, the weak ultraviolet radiation of more or less distant stars is replaced by that of the nearby Sun, and especially by the powerful ionizing effect of lightning. Furthermore, the density of molecules in the atmosphere is enormously higher than that in space—several billion billion (10^{18}) rather than several thousand molecules per cubic centimeter. This increase in density significantly augments the rate of encounters and reactions. And the temperature is many tens of degrees higher than it is in space, going as high as a few hundred degrees. But the biggest difference is the presence of the vast surface area of the oceans lying under the atmosphere. Giant waves, raised up by cyclones, mix the new molecules into the waters. There they find adequate protection against ionizing rays. A thousand times denser than air, the ocean is now a huge test tube in which molecules meet and combine.

The Primal Oceanic Soup

What do we find in this ocean? What new species have been created in this vast photochemical enterprise? To learn, scientists have tried to produce the same conditions in the laboratory. They put liquid water and the simplest gases known to exist in the primitive atmosphere into glass vials, which are then hermetically sealed. They submit this mixture, for days at a time, to continuous electric discharges from electrodes attached to the inside of the container. The water at the bottom of the vial is carefully observed. It first becomes cloudy, then yellow, and finally a yellowish brown. The liquid drawn from the vial has a disagreeable, pharmaceutical odor. When it is analyzed, alcohols, sugars, fats, and amino acids are found. Chemists call these "organic" substances. This terminology dates from the nineteenth century, when it was believed that such substances could be formed only by living beings. Yet here we are creating them in a test tube, without the help of glands or organs! (This experiment was carried out for the first time by the American chemists Stanley Miller and Harold Urey in 1954. It has been redone, and the results have been verified by a number of chemists all over the world. In science, it is a requirement that the same experiment be carried out by several different groups before a result is accepted. We can never have total confidence in results obtained in just one experiment, no matter what the quality of the work or the reputation of the experimenters. Too many uncontrollable factors, physical and psychological, can intervene. This mistrust is founded on many harsh memories. We've been "had" too often. In the search for truth, one can never take too many precautions.)

The brown swells of the primitive ocean, which in the distant past broke on all the planet's shores, also carried these precious, newly formed molecules. The winds carried their foul emanations over all the continents. But no living being had yet come to smell them, or to be nourished by this ocean overflowing with nutritive elements. Now the deluge is finished. The cloud layer thins out, and the Sun's rays begin to reach the ground. The Earth viewed from space is not blue, as it is today, but brown, the color of the oceanic soup, which has reached its highest level. Molecular activity is still going on. The game of combination and dissociation continues unceasingly in this fertile watery medium. New factors now enter the scene, factors that are destined to become dominant in the arena of growing complexity. We shall describe them in order.

Growth

Among the molecules that have just been generated, some possess "hooks" on each of their extremities, like the cars of a railroad train. They can join to form interminable chains called "polymers." As in the case of crystals, we see endless repetitions of a single motif. The difference is that the crystalline structure is rigid in three dimensions, while the polymer has the flexibility of a chain. Its volume is not fixed, for it has the ability to twist and even to coil back onto itself. This ability to take on a multitude of configurations, which we encounter here for the first time in nature, will play a crucial role in the elaboration

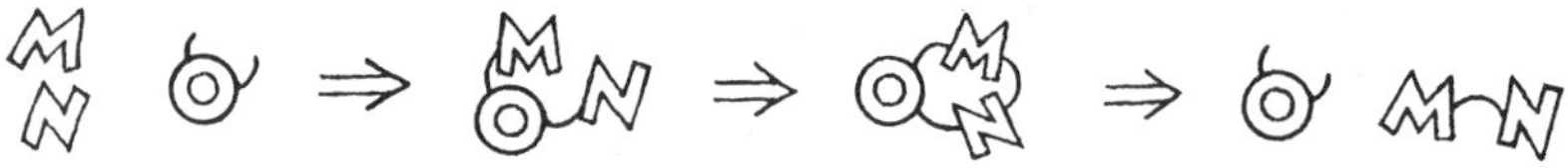

Figure 34
Catalysis.

of life. Let us note, in particular, the possibility of closing back on itself to form a ring. If other molecules are added, the ring can be transformed into a hollow sphere. Thus we see appear that fundamental notion of physiology, the "internal environment."

Catalysis

The joining together of individuals for a common purpose is not a novel event for us. In the core of hot stars, we have already seen protons link to carbon nuclei to accelerate their fusion into helium nuclei. This is the chain of events that regulates the generation of energy deep within Sirius, for example. In technical terms, carbon plays the role of a catalyst; it takes part in the reaction, but at the end it is unchanged, ready to begin again.

This phenomenon of catalysis, at the level of chemical reactions, occurs on a large scale in the primitive ocean. Let us look a little more closely at how it works (figure 34). Two molecules, which we call M and N, wish to combine to form a new molecule MN. Unfortunately, in their vast watery environment, M and N are rare. There is little chance of their meeting. But there exists a third molecule, O, that has two hooks with which it can capture samples of M and of N at the same time. These two isolated species find themselves side by side. They recognize each other, combine, and finally leave O to go away and live out their lives. Like a good servant, O is ready to begin again. It can thus considerably increase the rate of formation of MN.

Autocatalysis Foreshadows Reproduction

The couple composed of Mr. and Ms. Dupont are visited by Ms. Park and Mr. Duke. Ms. Park, who is unmarried, is an old acquaintance of Mr. Dupont, while Mr. Duke, who is also single, is a friend of Ms. Dupont from their university days. Ms. Park and Mr. Duke, who are both quite sociable by nature, have suffered greatly because of their bachelorhood and have long searched, without success, for soulmates. In despair they have asked for help from the Duponts, who have arranged a dinner for them. Everything goes well. The two guests are obviously quite pleased with each other. They decide to see each other again, and then to live together. A new couple is born. Surely you have recognized here the elements of catalysis. The Duponts play the role of catalyst in bringing together Ms. Park and Mr. Duke. More than that, we can speak of "autocatalysis" in a certain sense. It is one *couple*, in effect, who have accelerated the formation of another *couple*.

By this example I have hoped to illustrate the notion of autocatalysis, which plays a key role in evolution. The first synthesis of a particular molecule may occur very slowly deep in the oceanic soup. But if by chance this molecule possesses the property of autocatalysis—if it can serve as an agent for the formation of a molecule identical to itself— it will soon give birth to a companion. Then the two will go to work and there will be four, eight, sixteen, and onwards. We then witness a veritable demographic explosion that could take on catastrophic proportions. This language has a familiar ring to it. We speak of mice, rabbits, or human beings in the same way. In a sense, autocatalysis is a form of reproduction. It is undoubtedly more rudimentary, yet it contains all the essential elements. There is the production of a new system, as in simple catalysis, and there is "reproduction," in the sense that the created system is identical to the original system.

Feeding

Thanks to the long lightning strokes of the primal storm, the oceanic soup becomes filled with sugars and alcohols, substances that are particularly rich in energy. In parallel, other molecules are formed that have the capacity to capture and break apart alcohols or sugars while "draining" them of their energy. This is the beginning of pre- dation or of feeding, one of the principal activities of living beings. What happens to the energy acquired by the predatory molecules? There are many possibilities. Perhaps it will break them into simpler molecules, which will have lost the property of predation. In this case the process is a failure. But there is also the possibility that new combinations will be facilitated that will give birth to new properties. Here we see for the first time a type of behavior that will play a fundamental role in evolution: Failures are eliminated, while successes persist and open the way to new adventures.

Let us return for a moment to the molecular systems that have hollow cavities in their centers. Let us suppose that they use their atomic hooks to join with some of the fatty molecules that swim in the medium. One property of fat is that it repels water. It is "hydrophobic." It is the fatty oil in our skin that makes us watertight, and it is oil that keeps a duck's feathers dry. Here fatty molecules are responsible for keeping our hollow but now coated structure impermeable to water. Now we have a truly isolated system that is ready for an autonomous life, sheltered by its "membrane."

Some membranes possess the useful characteristic of allowing certain molecules, but not others, to pass through them. We say that they are "semipermeable." Let us now suppose that a molecule capable of breaking down sugars and liberating their energy (an "enzyme") is embedded in the interior of the membrane. We then see sugars penetrating the membrane and depositing their energy inside, after which their residue is seen to exit. This is an early version of digestion. I would emphasize that this scheme is not pure speculation; it has already been realized in the laboratory.

In short, the great functions of life—growth, reproduction, and feeding—already exist, foreshadowed, deep in the primal oceanic soup.

The First Energy Crisis

> Threatened by famine, the growing complexity survives and develops, thanks to solar energy.

At the start, the functions of nutrition and reproduction (through autocatalysis) are very rudimentary. They are not necessarily produced in the interiors of the same molecules. They develop, here or there, with a greater or lesser degree of success, by utilizing the vast reserves of energy that have accumulated in the ocean. For millennia the process of organization continues without rest. From our observatory beyond time and space, we follow its progress. In year X, the first molecule with a certain property appears in the broth; in year Y, a system of more than a thousand atoms is observed for the first time. With the multiplication of energy-consuming systems, though, the oceanic reserves start becoming depleted. So great is the appetite for energy that even the last remaining alcohol molecules are fought over. This crisis could conclude with a general famine and the destruction of all the complex molecules that have been so laboriously constructed. The progress of organization is thus seriously threatened.

The Sun shines steadily on, but its energy has no effect in dealing with this crisis until there appears a special molecule, a rudimentary ancestor of chlorophyll. Through a mechanism that foreshadows photosynthesis, it can capture solar photons and store their energy. It is the savior of the starving systems, provided they can find a way to collaborate with this precious partner. The first energy crisis is resolved through the development of solar energy, which will henceforth be the animating force for all forms of animal and vegetable life (*N12*).

9

Images of Biological Evolution

The Machinery of the Cell

> Midway between the atom and the animal, the cell already performs all
> the great functions of life.

I am not a biologist. I do not have the necessary depth of knowledge
to offer a comprehensive discourse on Darwinian evolution. I shall
not, however, resist the urge to call attention to certain high points.

The cell is the basic element, the foundation, of all living things.
We are assemblages of cells. Our bodies contain more than a hundred
thousand billion (10^{14}) of them, tuned harmoniously to one another.
These cells are not all identical; there are around two hundred varieties
in the human body. Each variety plays a set role. Some form the
bones, others the hair. Still others swim in globules through the blood.
Their dimensions, generally, are measured in microns (thousandths
of a millimeter). Their shapes can be spherical, cylindrical, branching,
or otherwise, according to their assigned function (figure 35). An average
cell contains around a thousand billion (10^{12}) atoms, similar to the
number in an interstellar dust particle; but there is a major difference
between the two. In the celestial dust grain, organization is minimal.
A simple motif is repeated over and over: one oxygen, one magnesium,
one silicon, one iron, then another oxygen, another magnesium, and
so forth. Furthermore, the atoms are fixed in place by rigid bounds
and can be dislodged only by the melting of the entire crystal. By
contrast, entire volumes, full libraries, are needed to describe what
we already know about the complexity of cells (figure 36). And yet
we know only an infinitesimal fraction of their secrets, of their structure,
of their behavior.

All cells share a certain number of common elements. First there
is the nucleus, in which the "genes" are stored as if in a safe. The

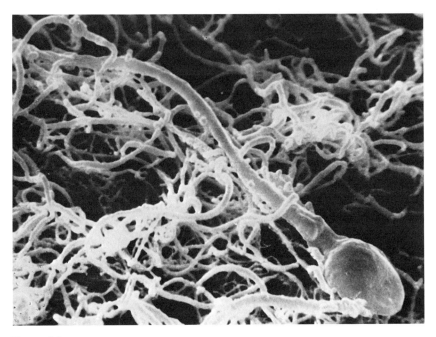

Figure 35
A human spermatozoon. Despite its tiny size (several hundredths of a millimeter), this cell contains all the paternal hereditary information. (Histology laboratory, Bicêtre Hospital)

"genetic code," which contains all the information necessary for living and for reproduction, is inscribed in the genes. The information is encoded by means of a special alphabet in which the "letters" are molecules (A3). The sequence of these molecules forms a substance called DNA (deoxyribonucleic acid). We do not know when DNA first appeared on the Earth, but it was probably in the primitive ocean. Since that time DNA molecules have been faithfully transmitted at the moment of reproduction. They enjoy a kind of immortality. They will last as long as life on the Earth (and perhaps longer if interstellar travel becomes a reality).

The nucleus is surrounded by a gelatinous substance called "cytoplasm." This substance is surrounded in turn by a membrane that defines the cell's "interior world." The cytoplasm is populated by a multitude of small entities, called "organelles," with well-defined functions as the cell's "lungs," "liver," and so forth. The lungs, for example, are the "mitochondria." Oxygen absorbed by normal respiration is delivered to the cellular membrane through the circulation of the blood. It penetrates the cell and combines with the mitochondria,

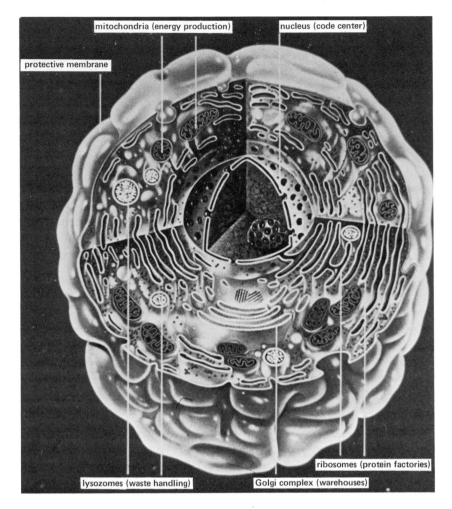

Figure 36

Sketch of a cell. At the center is the nucleus, which stores the molecules of DNA that carry the genetic code. The mitochondria produce energy; and the ribosomes make proteins, which are then stocked in "warehouses." The lysozomes eliminate waste products. (British Museum) which use it to release energy. The energy is then stored by special molecules called ATP (adenosine triphosphate), which are the cell's "rechargeable batteries." Later they will travel to places where vital functions of the cell require energy and will relinquish their store. Cellular respiration is not only a precursor of pulmonary respiration, it *is* respiration. The in-and-out motion of our lungs is merely a means of getting air to the cells.

Another class of organelles, the "ribosomes," have for their mission the building of the proteins needed for animal life. These are true assembly lines. Nutritive elements enter the body by means of the mouth. Then, prepared by the digestive system, they pass through the cell membrane and reach the ribosomes in the form of amino acids. According to plans furnished by the genetic code, the amino acids are hooked onto one another to construct proteins. There is some waste: Expelled by the cell and carried away by the blood, the waste materials are filtered out and evacuated in the urine. As in the case of respiration, then, the ultimate digestion occurs in the cell.

In the cells of vegetable matter there is a colored organelle called a "chloroplast" that is responsible for energy production. Plants absorb water through their roots in the soil and carbon dioxide from the air through their leaves. These two substances penetrate the cell membrane and meet in the chloroplast where, with the aid of sunlight, they combine to form sugars. This is the operation called "photosynthesis." The sugars, stored in the form of starches, can be used in turn to nourish animals that feed on the plants. Oxygen is the "refuse" of photosynthesis. Evacuated by the plant, this gas becomes part of the atmosphere. Plants alone are responsible for the presence of free oxygen in our atmosphere. No other planet in the solar system has any.

The reproduction of plants and animals is based on the reproduction of cells. First the molecules of DNA prepare exact copies of themselves. These two twins then separate and migrate to opposite sides of the cell. The cell divides, and the membranes close up to form two identical cells, ready to begin again. Life for each of us began with a single cell: the maternal egg fertilized by the paternal sperm (figure 35). This cell divides into two, then four, then eight, until it reaches the hundreds of thousands of billions (10^{14}) of cells in the adult. At the time of sexual encounter the process begins again.

The Origin of Cells

How did a system as evolved and as efficient as the cell come into being? Truthfully, we know little about it. One fascinating theory would make it the result of a collaboration. Very simple systems might have found it advantageous to live together and pool their talents in a form of symbiosis.

On our planet, the oldest known rocks are situated in Greenland. A sedimentary deposit has been found that was laid down 3.8 billion years ago, less than a billion years after the formation of the Earth (*N13*). At that time the power of the initial volcanism was waning. The first oceans, almost boiling hot, abounded in complex molecules created during the great deluge. In a recent discovery of the utmost importance, a vast population of microfossils was found in this sediment. Blue algae, among others, were identified. These are microscopic single-celled organisms capable of carrying out photosynthesis. They can be found today in the hot water that erupts from geysers in Iceland. They thrive at temperatures in the neighborhood of one hundred degrees centigrade. Under the microscope, the cells of these algae are quite disconcerting. They do not have a nucleus or any of the other usual components of a cell. They are composed simply of a gelatinous mass closed off by a membrane. Moreover, in the desert terrain that surrounds these geysers, we find vast populations of bacteria—organisms composed of only a single cell, also without apparent internal components.

According to the theory of collaboration, these are the simple cells that one day associated to form the complex cells of living beings. Each primitive organism became a particular organelle. The bacteria have, as mitochondria, taken responsibility for cellular respiration. In plants, the blue algae have, as chloroplasts, been assigned to photosynthesis. To bring together entities that already exist in order to form a more complex and more efficient object is indeed one of the favorite recipes of nature in the state of gestation.

The Great Darwinian Tree

In 4 billion years we passed from blue algae to human beings.

Again, despite my desire, I cannot properly guide you along the paths of biological evolution. Fortunately, others have marked out the course with great dexterity (*N14*). I would, however, like to salute the great moments in natural organization, in keeping with the spirit of the preceding chapters. In light of our actual knowledge, let us try to highlight the most evolved being at each stage in the Earth's history.

Bacteria and blue algae seem to have held the ascendancy for 3 billion years. The oldest multicellular organisms, to our knowledge, are the jellyfish, which appeared 700 million years ago. Surely there are older ones, but their remains are difficult to identify. One hundred million years later come the first crustaceans, which possess an external skeleton that leaves a fossil record. In another hundred million years (500 million years ago) the skeleton moves to the interior; the reign of fish has begun.

Life until now has been exclusively confined to the ocean and to lakes. The egress from the water takes place 350 million years ago. Thanks to the ozone layer, the atmosphere is now protected against the lethal rays coming from space. This ozone layer is itself the product of the respiration of aquatic vegetation in preceding eras. This is the beginning of the epoch of reptiles and birds. Mammals appear a little later but do not really proliferate until the dinosaurs have disappeared, around 63 million years ago. Among the mammals, a species of small shrew, which came into existence around 60 million years ago, carries the promise of the human brain in its genes. From this lineage come the various branches of the monkey family (figure 37), and from one of these branches are descended the hominids and the first men. The human body is made up of about one hundred billion billion billion (10^{29}) elementary particles. It is the particular arrangement of all these particles that allows you to concentrate your attention on the pages of this book.

A Catastrophe of Planetary Scope

A major event occurred on the Earth around 63 million years ago. Entire races of plants and animals disappeared. Species as different as dinosaurs, ammonites, and giant ferns were stricken from the ranks of the living. What happened? Recent evidence indicates that the cause of the mass slaughter was most probably astronomical in nature.

Chemical studies reveal abnormally high levels of certain rare metals, such as iridium, osmium, and gold, in the geological strata of this period. That is, they are abnormal compared to what is usually found on the surface of the Earth, although—and this is the crucial point—the relative quantities of these metals are in good accord with the abundances measured in fallen meteorites. It is as if a massive shower of meteoritic material fell onto the Earth during this epoch.

Let us imagine a meteorite many kilometers in diameter hitting our planet. Under the force of its impact, it volatilizes. Clouds of rocky dust are carried upward to the limits of the terrestrial atmosphere. Massive volcanic eruptions (like that of Krakatoa in 1883, for example) can darken the sky in a similar way. But how can such dust clouds

Figure 37
The most advanced steps in cosmic evolution. The monkey is one of our cousins. About 15 million years ago, its ancestral line branched to give the primates and man. Through the eyes of the young girl, the universe becomes aware of itself. The acquisition of intelligence involves the cooperation of about 10^{29} elementary particles. (Photo Jacques Véry)

bring about a mass slaughter? We do not know. Did the prolonged reduction in solar flux interrupt the life cycles of vegetation? Why did some animals survive and others die? Moreover, was the slaughter necessarily due to a meteorite? Perhaps the solar system encountered a dense interstellar cloud during this period, like the ones we see in Orion. Such a cloud could have produced an avalanche of interstellar dust (equally rich in iridium, osmium, and gold) onto the surface of the Earth. The important fact, though, is that the arrival of the layer of extraterrestrial material coincided with the massive extinction of individuals and species of animals over our entire planet.

This event appreciably altered the course of evolution of terrestrial life. During the previous 200 million years, the saurians were the most important branch of the animal kingdom. Mammals had existed for a long time, but they were a very timid form of life—minuscule beasts, about the size of our rodents. Their numbers were small and their development very slow. After the disappearance of the giant reptiles, everything changed. The population of mammals grew quickly, and their development accelerated. In a few tens of millions of years they attained the level of the monkey, other primates, and man. Coexistence with the saurians was apparently not very beneficial to the first mammals. If this is true, then the falling of the celestial rock takes on considerable importance in our history of the organization of matter. It brought about the elimination of the "obstacle" posed by the presence of the dinosaurs, and thus renewed the progress of complexity.

Life Involves All Levels of Reality

A woman is resting on an unmade bed. The scene exudes calmness and tranquility. But deep inside her body another scene unfolds. There is tumult, chaos, merciless strife. Billions of spermatozoa ascend to attack the single ovum that can assure them of survival. The competition is relentless; all of the combatants but one (or perhaps two) will perish here in a few minutes in a frightful mass slaughter. Beneath the cell membrane of each of these spermatozoa is yet another situation. Chemical reactions that produce proteins continue in their usual rhythm. Atoms join or dissociate, unconscious of the tumult on the level above them, or of the peacefulness being enjoyed on the next higher level. The nucleons of the nuclei witness as passive spectators the game of the electrons as they bring about molecular combinations, just as at a still deeper level quarks are powerfully bound in the hearts of the nuclei.

All the levels of the real world—past, present, and future—come together in this scene of everyday life. Quarks were bound into nucleons in the original soup of the first microseconds of the universe. Nucleons joined to form nuclei in the fertile environment provided by stars many billions of years ago. The genetic code, inscribed in the sex cells, was built up in the primitive ocean. Today the action continues at the levels of chemistry, cellular activity, and the sentimental life of lovers. It carries with it the promise of its own propagation into the future. "Life" is represented at all these levels; the simplest act has its roots at the beginning of time.

The Chemical Elements of Life

> In the primal soup, elements were tried out and chosen on the basis of their merits.

Let us go back to the early history of our planet. In the terrestrial crust, in the atmosphere, and in the primitive ocean are found some eighty stable chemical elements, each with well-defined properties (*A3*). Their abundances differ greatly. Like a child playing or a mason testing stones, nature "tries out" these atoms. Through chance combinations, sometimes successful, sometimes not, they are assigned to particular roles. Some will be needed in quantity, like calcium in the bones. Others will occupy key positions but will require only infinitesimal amounts, like iodine in the thyroid. Some elements with particularly diverse aptitudes will take on many roles in the unfolding of vital processes.

The predominance of hydrogen and oxygen recalls the crucial role played by the dissolving power of water. Using carbon atoms, the molecules associated with the genetic code take form: amino acids, proteins, nucleotide bases (*A3*). Combining with hydrogen, oxygen, and nitrogen, carbon gives rise to an almost infinite variety of structures capable of storing information. Sulfur also plays an important role here. Blood uses iron to carry oxygen from the lungs to the cells, but researchers have discovered that an atom of iron cannot become part of hemoglobin unless a certain protein based on copper is present. How did nature make this discovery? Among all the molecules that could collect energy and release it at the proper moment, nothing surpasses the efficacy of ATP, which is built around phosphorous. Nature chose it to ensure the digestion of nutrients on the cellular level. Life is based on chemistry, and chemistry is based on the exchange of electrons. Six principal elements participate in such exchanges: chlorine, sulfur, and phosphorous as donors of electrons; magnesium, sodium, and potassium as receptors.

With the enzymes we find once again the phenomenon of catalysis, which we have already encountered in the stars and in space. These highly specialized molecules intervene at specific instants in the cycle of life. In many cases it is the presence of a particular atom in their architecture that creates their specificity. Zinc participates in the digestion of alcohol and proteins. Manganese helps in the formation of urea, and cobalt in the synthesis of the genetic code. Copper, as we have seen, assists in the incorporation of iron into blood, and it also plays a role in the pigmentation of skin, in the elasticity of the lining of the aorta, and in photosynthesis. It has been shown that living organisms also use (albeit in very small amounts) boron, fluorine, silicon, vanadium, chromium, selenium, and tin. At least 27 elements come together in the machinery of life. Biologists are working today to extend this list (*N15*).

The noble gases (helium, neon, argon, krypton, xenon) are absent from the list. We might have expected this. The rigidity of their bonds disqualifies them from the game of combinations. Likewise silicon, a close relative of carbon but one hundred times more abundant in the crust of the Earth, only participates in infinitesimal proportions. Its great fault is that it is not soluble in water.

10

Life beyond the Earth

Life in the Solar System

Are there plants and animals on other planets of the solar system? In the nineteenth century people spoke freely of Venusians or of Martians and their grand canals. Since the beginning of space exploration, however, our belief has ebbed.

The Moon and Mercury do not have atmospheres. Why? They are not massive enough to retain a gaseous envelope. At their birth they contained vast quantities of gas in solution in the hot rock, just as did the Earth. This gas, expelled through many volcanic vents, escaped into space, leaving the bare ground below without protection. Continually bombarded by energetic particles from the Sun and elsewhere, the rocks crumbled. A layer of fine dust particles accumulated over the course of the ages. The astronauts have left their footsteps in that layer of dust on the Moon. Mercury and the Moon are arid and desertlike (figures 27 and 28).

For quite different reasons, Venus, with its very dense atmosphere, does not present more favorable conditions for life. The atmospheric mass of carbon dioxide creates a planetwide "greenhouse" effect. It traps the heat of the Sun and warms the surface of the planet to nearly 500 degrees centigrade. There are no liquids. In this case it is the extremely high temperature that discourages molecular bonding.

And Mars? In 1976, NASA sent out two unmanned Viking probes, equipped with fully automatic chemical laboratories. They sat on the red soil, grabbing onto specimens, which they crushed and analyzed. How do we decide whether a planet possesses animals or plants? There is a simple method: Look for characteristic molecules. On the Earth, even in the most arid deserts, there are always myriads of organic molecules. The wind carries microscopic pollen and spores

Figure 38
Callisto, a satellite of Jupiter, contains enormous amounts of water in the form of
ice. (NASA)

everywhere. In the Martian soil, however, there are far fewer complex
molecules than are found deep within the glaciers of Antarctica. We
have little hope of finding life, even of the most primitive sort, on
Mars (figure 30).

And what about other places? Some of the satellites of Jupiter and
Saturn possess atmospheres (figure 38). They could conceivably host
certain plant forms. One important discovery described in the next
section gives weight to this hope.

Amino Acids in Meteorites

The early steps of life are found imprinted on the debris of shattered
planets.

Meteorites are rocks that fall from the sky. Each year they arrive by
the hundreds, over the entire surface of the Earth. Their dimensions
range from a few centimeters to several meters. On entering the
atmosphere they produce a long trail of light (like shooting stars), often
accompanied by a clap of thunder. They hit the surface violently and

often penetrate it. Where do they come from? Before they meet up with the Earth, they silently circle the Sun, like any planet. A few meteorites have been photographed just as they were entering the atmosphere, and their orbits have been retraced. Most come from beyond Mars. For the reader who has never seen a meteorite, I recommend a visit to a museum of science. To touch the polished surface of a meteorite that has spent many years roaming among the planets of the solar system can make you dizzy, just like gazing at the Milky Way on a dark summer night.

There are two basic kinds of meteorites: stony and iron. Iron meteorites are very dense and are a rusty metallic brown in color. Their surfaces are often pocked by deep cavities, caused by friction during their descent through the Earth's atmosphere. Space capsules receive similar damage when they reenter the atmosphere.

Stony meteorites are grayish, like pebbles in a field. Some contain small glass spheres called "chondrules" embedded in their matrix. These chondritic meteorites sometimes contain significant quantities of crystallized water and carbon; these are called "carbonaceous chondrites." When analyzed, the carbonaceous material is found to contain hydrocarbons (such as tar and oil) and even amino acids! Was this organic matter always present in the meteorite, or is it a contamination acquired deep within the terrestrial atmosphere? These kinds of molecules are abundant in air and rainwater. How can we tell, then, whether the meteorite has not acquired them *after* its arrival on the Earth's surface? It is a thorny question that has stirred up long, heated, and even venomous discussions. But today we know that amino acids do in fact exist in meteorites before they enter the atmosphere. In the next section I shall explain how, thanks to the work of Pasteur, we have been able to settle this question.

Pasteur and the Sugars

Sugar can be made in two distinct ways. It can be extracted from plants such as sugar beets or sugarcane, or it can be made in a laboratory using water and carbon dioxide. Are the sugars derived from these two processes the same? There is a difference, but not one you would be aware of while sipping your morning coffee. The difference is revealed by shining a beam of light through a sugar solution. The sugar derived from a living source reacts in a precise fashion, while the sugar produced in the laboratory shows no reaction. (More precisely, we use "polarized" light. The natural sugar will make the plane of polarization turn, but the laboratory sugar will not. This rotation can be observed by means of polarized glasses.) What is the cause of this difference?

There are, in fact, two types of sugar molecules. They contain the same atoms (carbon, oxygen, hydrogen), but their geometrical structures differ. Conjure up in your imagination a spiral staircase. It can turn in one direction or the other. In the same way, some molecules turn in one direction, and some turn in the other. They are otherwise identical, like a left hand and a right hand in a mirror. Pasteur showed that laboratory sugars contain both types, whereas those made from a live source contain only one. (A solution consisting of a single type of molecule would make the plane of polarization turn in one direction. The other type would make it turn in the opposite direction. When the two types coexist, their effects cancel out, and the plane of polarization remains fixed.) In the realm of living things, the sugar-making molds are such that they can make only the type of sugar out of which they themselves are made. There seems to be a kind of selection going on here, similar to the kind we encountered in the formation of a crystal. The fact that all plant and animal life manufactures the same type of sugar is a profound manifestation of the great unity of living things on the Earth.

In the beginning, in the oceanic soup, the situation could have been different. There is no reason why the lightning bolts of the initial deluge should have favored one sugar over the other. Later, however, one of the varieties disappeared. Why? This is a much debated question (N16). Perhaps at the appearance of the first systems capable of eating and reproducing, the two populations had to devour each other (both sugars can serve universally as nourishment). One population must have eliminated the other. On another planet, the winning population could have been different. This situation (two possible varieties, a single survivor) is not restricted to sugars. A large number of complex molecules have undergone this process. Among terrestrial beings, for example, only one form of amino acids exists. The other is missing. But both forms exist in carbonaceous chondrites. This is the proof that atmospheric contamination does not account for the amino acids found in meteorites. They can only be due to an indigenous source, and thus they are formed somewhere other than our planet.

Shattered Planets

What do we know about the early life of meteorites? We have good reason to believe that most of them come from much larger "parent" bodies. These would be planetesimals like Phobos (figure 39), whose

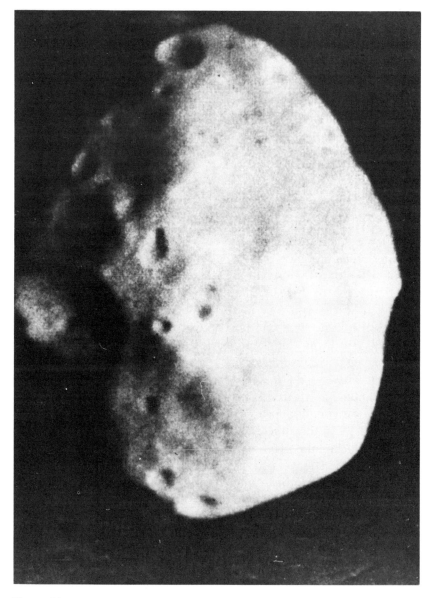

Figure 39
Phobos, a satellite of Mars. Meteorites apparently come from small planetary bodies that have broken into pieces due to collisions with other solar system bodies. (NASA)

diameters do not exceed a few hundred kilometers. In the interiors of these planetesimals, as in the Earth or the Moon, matter is differentiated by density. The heaviest metallic substances will have flowed toward the center, while the much lighter rocky material will be found in overlying layers. Collisions with other solid bodies can break these planetesimals apart. Their fragments become stony or iron meteorites, depending on their source.

The amino acids of our carbonaceous chondrites may have been formed in parent bodies that no longer exist today. The presence of water in these chondrites should not surprise us, since we know how important this substance is in the production of terrestrial life. Let us note in passing that Callisto, a satellite of Jupiter, seems to consist largely of frozen water. We should not abandon all hope. The simultaneous presence of two varieties of amino acids in certain meteorites has shown us that these molecules existed before their arrival on Earth. Their very presence teaches us something else. If a single variety of a molecule still exists in the terrestrial biosphere, both varieties surely coexisted in the primal soup. We conclude that, on the shattered planetesimals, the organization of matter reached the stage at which amino acids have formed, but not the stage at which one variety has eliminated the other through competition.

The discovery of amino acids in carbonaceous chondrites, like the unanticipated observation of interstellar molecules, has deeply stirred the scientific community. It once again illustrates the astonishing fertility of matter.

Life in the Universe

There are probably millions of inhabited planets. But they are very secretive.

And elsewhere? Beyond the solar system, among the billions of stars that make up our galaxy, among the billions of galaxies that make up our universe, are there living beings? The stars are far away. Explorations at these distances are projects for the millennia to come.

In place of going to see for ourselves, we can make observations and search for proofs of one sort or another. We can ask ourselves, for example, if other stars possess a retinue of planets like our Sun's. A planet, as we have seen, presents an ideal solution to the multiple problems of matter in the process of organizing itself. We know in any case that single stars are in the minority. More than half of all stars live with one or more partners. It seems quite plausible that some of these celestial bodies have compositions similar to that of the Earth. Presumably some of their orbits are such that they receive the amount of heat necessary for the development of life. The number of inhabited planets could thus be very large. Some authors speak of millions in our galaxy alone. This estimate is very uncertain, to be sure. To my mind, it simply reflects our realization of the relentlessness with which life develops everywhere when conditions permit—and of its ability to alter conditions to improve its chances of progress.

The Sun was born quite late in the life of our galaxy. Many billions of stars were born before it. What beasts abound on the surfaces of their hypothetical planets? Jellyfish, dinosaurs, hominids, or something entirely different? Planets by the thousands may already have attained technologies superior to our own and may be communicating with one another through radio messages or interstellar voyages. We should be able to intercept these messages. Our radio telescopes are sufficiently powerful to receive the equivalent of Voice of America or BBC emitted from a distance of several light-years. Many tentative attempts at listening have been made. The best antennae of our planet have devoted time to this search, but without success. Perhaps you have tried to catch the emissions of distant countries on a short-wave radio. By turning the dial, you can sweep through the entire range of frequencies. More often than not, all you hear is "static"—an incoherent succession of hissings, roars, and shrill or deep notes. Then suddenly, feebly, a voice, a music, comes through. Even if the language is foreign, even if the distortions are overwhelming, you can still distinguish the programmed emissions from the noise. Up to the present all that we have heard from space has been static. Not a single signal has been received from an intelligent source, not a single "program" that shows any intention of communication. But we are far from having covered all the possibilities. Systematic exploration in all directions on all frequencies, using all possible radio bands, has hardly begun. It is much too early to despair.

And what about interstellar tourism? In this case the problem is not one of insufficient data! Recorded reports (occasionally backed up by local police authorities) of observations of "unidentified" flying objects would fill entire libraries. UFOs have been seen, photographed, sometimes tracked by radar. Spectators have been carried off on these objects. Some have disappeared forever. Others, like Marco Polo, have returned to tell of fantastic events. However, the situation is shrouded in confusion. Under critical analysis, the vast majority of testimonies dissolve. We find cases of fraud, hallucination, or simply the desire to make oneself interesting. The photographic documents are blurred, the radar echoes indistinct. No useful information concerning the aeronautics or mode of propulsion of these objects can be extracted from the reports. Nor do we have any credible information concerning their extraterrestrial origin. Some cases, to be sure, remain unexplained, mysterious, and troubling, and taken together they demand attention.

Let us try, all the same, to give our discussion an added dimension. Let us for a moment put on the skin of a prehistoric man, perhaps an inhabitant of the caves of Lascaux. His brain is already as developed as ours; the polychromatic drawings with which he decorates the walls of his caves offer ample proof of this. He is, however, ignorant of radio waves and communication at a distance. He lacks the millennia of technological developments that have transformed our perception of reality. The point is that there exist forces in nature that can escape our perceptions. Today we know how to manufacture television sets with which electromagnetic radiations become perceptible. Yet who would be bold enough to state that we know and perceive all forces, all waves, all means of communication?

Extraterrestrial civilizations are not inundating us with radio messages. Visitations to our planet remain very doubtful. But that fact does not end the discussion. The possibility of other types of travel and of other types of telecommunication remains with us.

11

The Future of the Earth

The Death of the Sun

Botanists strolling through a pine woods can mentally classify the trees by age groups. In their minds they carry an image of the entire life cycle of the pine tree. At their feet are the young sprouts, above their heads the vigorous adults, and lying on the ground are the decaying trunks of trees that have died. They could tell any one of these pines what its future will be. Astronomers are in a similar position with respect to the stars in the firmament. By identifying stars at various stages of their life cycles, they can see revealed, in a single sweeping panorama, the process of stellar birth, life, and death. There is, of course, one star in which we have a more than merely academic interest: our Sun. The fate of the human race depends intimately on what the future holds in store for the Sun.

In an earlier chapter I described the death of small stars. They pass successively through the red giant, planetary nebula, and white dwarf stages, before becoming black dwarfs. I shall now try to reconstruct the sequence of events in the latter life of our Sun as they might appear to a hypothetical extraterrestrial observer.

The Sun's hydrogen reserves promise us about 5 billion more years of tranquillity. The Sun will remain very nearly what it is today: a yellow star whose enormous disk seems, because of its great distance, to have about the same apparent size as the Moon. Upon the exhaustion of its central supply of hydrogen, the Sun will become a red giant and ignite the fusion of helium into carbon and oxygen. Betelgeuse (the left shoulder of Orion), Aldebaran (in Taurus), and Antares (the heart of Scorpio) belong to this class. We can see with the naked eye (and better yet with binoculars) that these stars are red. They truly are giants! If we mentally superimpose the center of Antares on that of the Sun, we find that the star swallows up not only the Sun but also Mercury, Venus, and the Earth!

When the Sun reaches this phase, its volume too will increase, and its surface temperature will slowly decrease. From yellow, its light will shift to orange and then to red. From the blue of the sky to the soft hues of dawn and dusk, all atmospheric phenomena will be profoundly affected. Will the Earth cool off? No, quite the contrary. The increase in the solar surface area will more than compensate for the decrease in temperature. The swollen red disk will send us more heat than we now get from our familiar Sun.

What will happen to our unfortunate planet at that time in the distant future? Do we have the basic knowledge to sketch out a reasonably realistic scenario, at least in broad outlines? I am not totally convinced, but let us try anyway (*A*5). As the heat increases, the polar ice caps begin to melt, steadily raising the level of the oceans and creating thick layers of clouds that, for a time, hide the Sun. These clouds largely erase the climatic contrast between poles and equator. An Amazonian jungle, hot and humid, covers the entire planet, and, as in a greenhouse, luxuriant vegetation flourishes everywhere. Then the atmosphere begins to evaporate into space. The skies become clear. Baked under the enormous red disk, dessicated vegetation spontaneously bursts into flames. Using up the remaining oxygen, endless brush fires consume all the organic matter on the surface of the Earth.

A lunar landscape emerges. On the continental blocks and in the depths of the basins from which the oceans have evaporated, the mineral kingdom regains the place it held during the planet's earliest age, a place that it never lost on the Moon (figure 28). After a few more hundreds of thousands of years the rock itself begins to melt, just as it does in volcanic vents today. In red-hot cascades, waves of incandescent lava flow down the mountains and collect in the ancient oceanic depths.

The monstrous red belly of the Sun continues its inexorable expansion, exuding from its bowels a powerful stellar wind. Under its impact, the inner planets—Mercury, Venus, Earth, and perhaps Mars as well—slowly evaporate. Their matter is swept along by this hurricane, driven into space in chaotic billows of vapor. Later on, the expulsion of matter takes on a more erratic and violent character. The outer planets—Jupiter, Saturn, Uranus, Neptune, and Pluto—are vaporized one after another under the impact of these torrid gusts.

Seen from afar, these events probably take on the rich coloration familiar to astronomers from observations of planetary nebulae (figure 22). A hot central blue-violet star is surrounded by concentric rings, ranging from yellow-green to red at the edges. The central star is the residual core of a dying red giant, while the rings are made of the stellar matter that has been hurled out. Greatly diluted, and stimulated by the light of the central star, these gaseous masses become transparent and glow. Oxygen atoms provide the greenish fringe, while atomic hydrogen and nitrogen furnish the red corona. The dying Sun, however, does not exhaust itself completely. A denuded core remains, a "white dwarf" like the one that orbits Sirius (figure 23). In this way the material of our vaporized planet is returned to the galactic gas out of which it emerged 4.6 billion years ago. From this dilute material new nebulae form, and in their midst new stars and new planetary systems are assembled.

Visions of eternity in perpetually repeating cycles. . . .

Reviving the Dying Sun

The best possible use of thermonuclear weapons.

At times I worry about the fate of our descendants who will be alive during this critical period of the Sun's death. Must they inevitably perish?

I see three possible ways out of this dilemma. First, they might migrate to planets more distant from the Sun. Two satellites of Jupiter, Ganymede and Callisto, hold large reserves of ice (figure 38). Warmed by the radiation of the immense red Sun, they might, if appropriately managed, be made habitable. Today we already know how to land people on the Moon and how to build undersea habitats. This solution is thus not utopian and may in fact be realizable in the not too distant future. My worry is that it will inevitably be reserved for a chosen few. And who will make that choice?

The second solution consists of moving the entire Earth to keep it at a safe distance from the menacing Sun. To accomplish this we would have to deploy batteries of appropriately aimed rockets on the Earth's surface, as we would to launch an artificial satellite. To obtain the required energy we must first have achieved the controlled fusion of hydrogen. I have calculated that, by "burning" about 10 percent of the water in the oceans, we could displace the Earth's orbit out to the present orbit of Saturn. There is a problem, of course, in that the level of the oceans would be lowered by about 200 meters. But we do what we must! These two possibilities (migration and changing the Earth's orbit) have one weak point in common: They are both short-term solutions. They would be useful only during the red giant phase

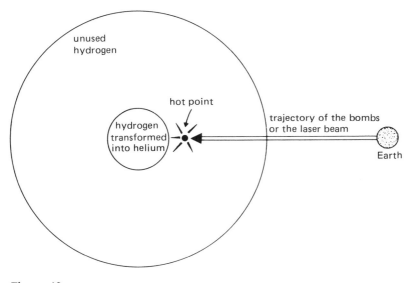

Figure 40
Rejuvenating the Sun.

of the Sun—a mere 100 million years. When the Sun becomes a planetary nebula, and then a white dwarf, the same problem will again arise.

Fortunately there is a third solution, much more difficult but also much more lasting. It is the revival of the Sun—in the same sense that we speak of reviving a failing heart (figure 40). Recall that the Sun obtains its energy by burning hydrogen into helium. The nuclear reactions responsible for this fusion take place where the temperature is highest, at the center of the Sun. About 50 percent of this central hydrogen has already been transformed into helium. In 5 billion years there will be no more hydrogen in this hot region. Then the Sun, deprived of its fuel, will enter the last phases of its existence. Yet there will remain vast masses of unburned hydrogen between the core and the solar surface. This is, in a sense, a malfunction in the machinery of the Sun. A "pump" is needed to circulate the fuel and to help rid the central furnace of the ashes of the fusion process. We could in this way prolong the life of the Sun from 10 billion to about 100 billion years!

For this project we must "stir" the material of the Sun periodically, much as one stirs a cup of coffee to mix the sugar and the liquid, or, even better, as one revives a campfire by pushing wood from the periphery into the hot coals at the center. To do this we must create a hot spot between the center and the surface, a little outside the fusion zone. I can see two possibilities. The first is to detonate super hydrogen bombs. With today's bombs we have already created temperatures much higher than those in the heart of the Sun. The problem is to get the bombs to their intended destination without their vaporizing along the way. Here I am fresh out of ideas. But, after all, we have plenty of time to think about it. It does not seem impossible to me that we may some day solve this problem. The second possibility is to aim a powerful, extremely concentrated laser beam at the solar surface. Here again, though, we must face the problem of assuring that the energy is not dissipated too soon.

A number of stars in the heavens seem to have continued their hydrogen burning well beyond what might normally have been expected. We are still seeking a "natural" explanation for this phenomenon. And what if it involves exactly such a voluntary intervention by the inhabitants of planets that depend on these stars for light and heat? Warned of their approaching fate, our distant cousins may have found the means to stir their star and thus prolong its life. One might object that these stars, called "blue stragglers," do not live long and give off dangerous amounts of ultraviolet radiation. As a result, there is little likelihood that they harbor any life like that on Earth. But, after all, we know so very little about the development of life. . . .

The image of a dying Sun preoccupied the Aztecs (*N17*). To keep it alive they periodically offered human sacrifices, young people in the prime of life slaughtered on altars atop their pyramids. Why not offer instead the tens of thousands of atom bombs stockpiled in our nuclear arsenals?

12

The Cemetery of the Côte des Neiges in the Constellation Orion

Spiral evolution unfolds through birth and death—a viewpoint familiar to Hinduism.

As a child I lived near a large cemetery in Montreal, an immense park planted with oaks and sugar maples. On the first days of spring, long before the tombstones were hidden by shrubbery, the crocuses emerged from the melting drifts of snow. I would go to see them in bloom, and I also watched the burials. The contrast between these events fascinated me. On one side were vigorous flowers and trees, thrusting their roots out from the moist banks of the graves; on the other, the polished wood coffins that men eased down into the dark holes. It was at once the beginning and the end, life and death—the ephemeral and the eternal intimately intertwined. This scene had no age; it simply was. It is not surprising that ancient civilizations deified the Earth, and that they combined in a single symbol the Earth and the womb, the two wellsprings of life. In those primitive times ritual ceremonies accompanied by sexual orgies announced the arrival of spring. New harvests were about to spring from the fertile Earth. Earth is Life. But it is also Death, the ultimate agent of dissolution of beings that have finished their existence. Decomposing in the soil, the formidable molecular machinery that makes up the smallest daisy or the humblest ant is torn apart. Cells are decomposed into complex molecules, which themselves are broken into simpler molecules, and so on.

This process of dissolution remains incomplete. Any fertile field contains vast amounts of organic matter. Each decomposing plant enriches the soil, making it more fertile, better able to nourish new sprouts. Birth, life, and death form the elements of a cycle that does not close upon itself. Each cycle makes some new contribution that influences the cycles yet to come. This is "spiral" evolution.

In this sense, the Earth is a primary material out of which both plant and animal life arises, and to which all life returns, only to reappear again. This is the wheel of life. The atoms and molecules that form our bodies have a long history. Many times in the past, life has borrowed them from Mother Earth. They have been the leaves of trees and the feathers of birds. In a few decades we shall no longer be here, but our atoms will always exist, pursuing elsewhere the complexification of the world.

Astronomy tells us that closely analogous events take place in the heavens. I rediscover my cemetery of the Côte des Neiges when I lift my eyes to the constellation Orion. It is in the winter, a little after sunset, that Orion appears in all its splendor: seven brilliant stars that the ancients associated with the hunter Orion, who was killed by Artemis (figure 41). The two shoulders are marked by Betelgeuse and Bellatrix, one foot by Rigel, the other leg by Saiph, and at the center are three stars that form Orion's Belt (sometimes called the Three Wise Men).

Radio telescopes have revealed the presence of two large interstellar clouds in this constellation (figure 42). They contain immense concentrations of matter, with a superficial resemblance to clouds in our own atmosphere. The dimensions of these nebular masses are measured in tens of light-years. They contain altogether as much matter as a hundred thousand Suns. The Sun in turn contains as much mass as two hundred thousand Earths. There are many such interstellar clouds in the sky, especially in the vicinity of the Milky Way, where they form extended dark regions (figure 43). The best known of these, the Coal Sack, is located in the southern hemisphere. It is visible to the naked eye; in fact, it looks like a photographic negative. Since it is opaque, it blocks the light of the stars located behind it, just as, at night, clouds in our atmosphere may locally hide the starry vault. These clouds (or nebulae) become visible when one or more very hot stars are embedded in their midst. Under the impact of stellar radiation, vast expanses of nebular matter begin to glow. This matter is colored a yellowish green, fringed with red and violet. The Trifid Nebula, for example, resembles a bouquet of anemones (figure 44). The Rosette Nebula has the delicate texture of a porcelain rose (figure 45). But to my taste the most beautiful is the Orion Nebula, located just a little below Orion's Belt (figures 46 and 47). On a very dark, clear night you can make it out with the naked eye, and with a pair of binoculars you can see it rather clearly. When someone asks me, "Of what use

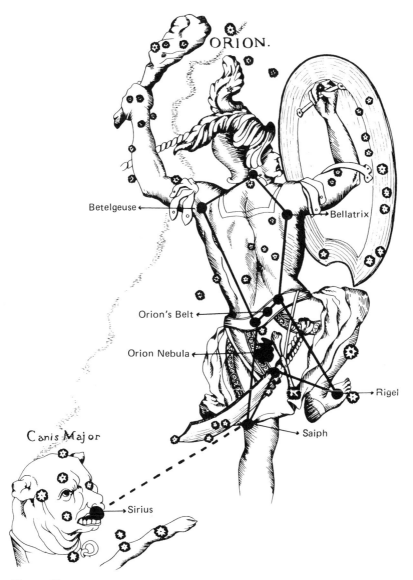

Figure 41
The constellations of Orion and Canis Major (after Vesalius).

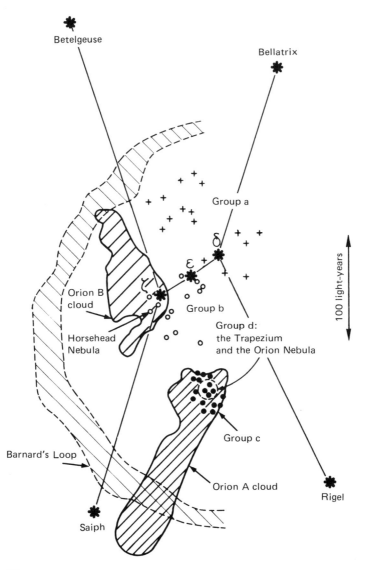

Figure 42

The constellation Orion. Asterisks mark the stars that delineate the constellation in the sky. The hatched areas are the interstellar clouds observed by radio telescope. The stars marked with a plus make up the oldest group (a; 8 million years). The stars of group b, marked with small open circles, are 5 million years old. Those of group c, marked with black dots, are 3 million years old. Group d, in which the Trapezium is located, is less than a million years old. The infrared cluster (less than 100 thousand years) and group d are located within the dashed circle. Barnard's Loop, which we see as the brightest feature, haloes the entire assemblage.

Figure 43
The Quill Nebula in Monoceros. In the midst of this enormous opaque mass, emblazoned with light, we can identify about twenty stars that were born in the last million years. That is after the appearance of the first humans on Earth. (Observatoire de Haute-Provence)

Figure 44
The Trifid Nebula. This is a great cloud of interstellar matter in which extremely hot stars trace out by fluorescence a red anemone. This cluster, located about 2000 light-years away, spans 10 light-years and contains about 300 times the mass of the Sun. (Lick Observatory)

is astronomy?" I answer, "If it had no use other than to reveal such beauty, it would have amply justified its existence."

I have already described the intimate relationship between heavenly bodies (stars and planets) and interstellar matter (gas and dust). I have related how stars are born of this material and how, in dying, they return their substance to it. Astronomical observations reveal, in the vicinity of the Orion Nebula, a major nursery of stars. Many of these stars are less than 10 million years old; they are contemporaries of the first hominids.

For us, Orion is a museum of stellar life. In the biology section of a science museum, for example, one can see series of rabbit embryos aged one hour, one day, one week, and so forth. Similarly, near the Orion Nebula we have identified five "litters" of stars of different ages. The youngest, still forming out of nebular matter, are less than a hundred thousand years old. At that time our ancestors were already working flint into tools. These young stars are truly stellar embryos. Not yet hot enough to emit visible light, they shine in the infrared.

Figure 45
The Rosette Nebula, a cloud of about 20 light-years radius, at a distance of 4000 light-years. In the darkest central region we see several stars born from this cloud. The black filaments that are silhouetted against the most brilliant part are opaque condensations from which stellar embryos are formed. (Observatoire de Haute-Provence)

It is apparently at this stage of a star's life that its retinue of planets begins to form.

Like fry in an aquarium, stars are born very close to each other. At the heart of the infrared cluster in Orion, the distances between stars are several light-weeks, while in the rest of the sky neighbors are usually several light-years apart. On hatching, fry live in close ranks; then, little by little, they detach themselves from the school, drift away, and finally take their chances and range throughout the aquarium. In the same way, stars slowly abandon their birth cluster and go to live their lives throughout the extent of the galaxy.

Quite close to the infrared cluster we find the Trapezium cluster. It owes its name to four blue supergiants, which are the stars that illuminate the Orion Nebula. There are also several dozen less luminous stars. The stars in this cluster, born together about a million years ago, are today hot enough to emit visible light. They represent the second developmental stage in our Orion Museum (figure 48). There are three more groups of young stars in the constellation, with ages of 3, 5, and 8 million years, respectively. The volumes of space occupied

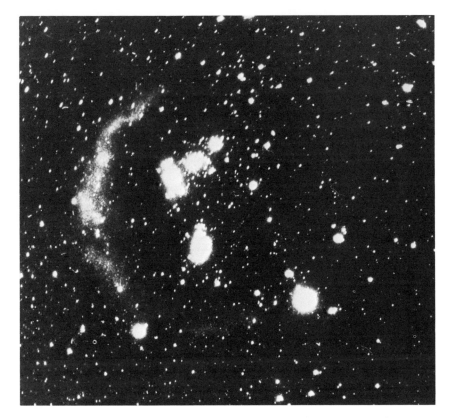

Figure 46

The lower part of the constellation Orion. Below Orion's Belt the Orion Nebula can be seen; this is a major center of stellar birth. About it is the faint arc of Barnard's Loop. The nebula apparently represents the collective effect of several supernovae that exploded tens of millions of years ago amid this sprinkling of stars. (Isobe, Tokyo Observatory)

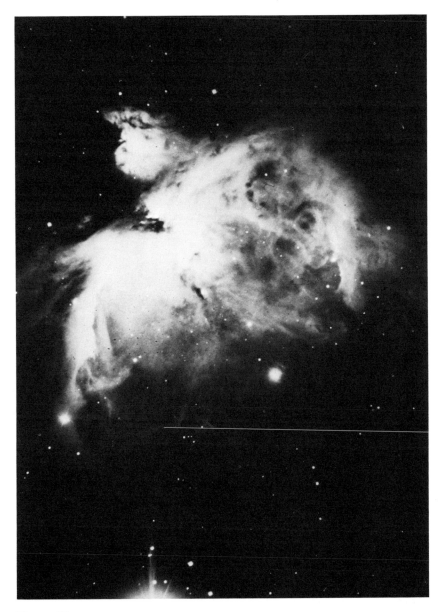

Figure 47

Just visible to the naked eye, the Orion Nebula presents one of the most beautiful sights of the heavens in a telescope. In the heart of the nebula four blue stars—forming the Trapezium—illuminate the materials dispersed in space out to very great distances. Atoms absorb this starlight and reemit in their own colors: oxygen in green light, hydrogen and nitrogen in red. Radio astronomical observations reveal that the Orion Nebula is but an infinitesimal part of the great "dark cloud complex" of Orion, like the visible part of an iceberg. The stars of the Trapezium were born from the contraction of this cloud, just as were a group of protostellar nebulae (detected by infrared telescopes) located behind the Orion Nebula. (Mount Wilson and Palomar Observatories)

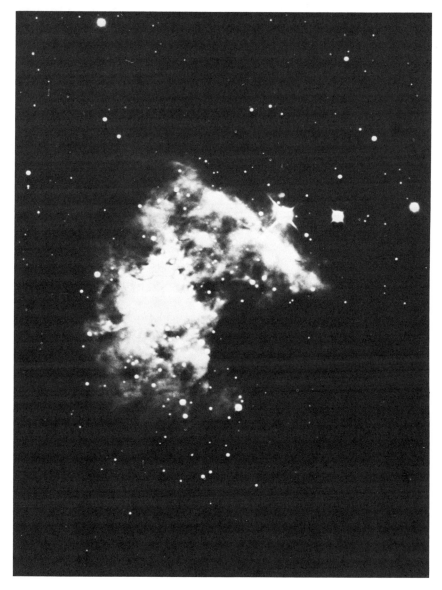

Figure 48
The Orion Nebula in the infrared. At this wavelength the nebula becomes partially transparent and allows us to see the four stars of the Trapezium. They are close to the center of the photo, haloed by diffuse light to their right. (Lick Observatory)

by these groups increase with their age. What we are seeing here is the dispersion of stars in the sky, just as, in our aquarium, the schools of fish eventually expand to occupy the entire accessible volume. After 12 million years the dispersal of stars is nearly complete. It becomes practically impossible to identify families of stars much older than this in the Orion nursery.

Twelve million years is, we note, more than the total lifetime of massive stars. Some stars therefore die before leaving the birth cluster. Like the supernova of July 1054, they explode and hurl their debris into space. This stellar material spreads out over tens of light-years into a vast network of ragged, turbulent gaseous filaments. We have counted more than a dozen stars that have ended their days in this nursery of stars. A luminous halo called Barnard's Loop circles Orion. Its center coincides closely with the position of the young clusters of stars. Quite in keeping with its appearance, it is presumably made of the accumulated debris from all these supernova remnants (figure 46).

I rediscover my cemetery of the Côte des Neiges in the constellation Orion. Birth and death, intimately intertwined.

Regions in which similar phenomena are unfolding exist in great numbers in the heavens. I have chosen Orion because it is one of the closest (1500 light-years away) and is easy to observe. Once someone wrote to me, "Since I heard you talk about Orion [on French television], I have looked at that constellation with a new eye, moved by the realization that stars are being born there before my eyes." Some 4.6 billion years ago, our Sun was born under similar circumstances. But where are the siblings of the Sun today? They are dispersed throughout the galaxy. We do not have any way to identify them.

Let us return again to these parallel cycles: *earth–living beings–earth* and *interstellar matter–stars–interstellar matter*. As I mentioned earlier, the sequence of these cycles is not a simple repetition. Stars have their own activity; they produce new atoms. These atoms fertilize space, making interstellar matter better able to produce new stars and new planets. The galaxy is not the same after the passage of a generation of stars as it was before. In the same way, the organic matter released by decaying plants serves as a fertilizer for new crops. On the Earth and in the heavens cyclic activities act as agents of increasing complexity. We find ourselves at the very heart of Hindu cosmology, where we rediscover Kali, the goddess mother, who is represented as being both the womb and the grave of all life. And here we also encounter the Hindu notion of "cyclic time," the source of spiral evolution.

13

Music from the Start

to Gilles Tremblay

Why Music Rather than Noise?

After the question, "Why something rather than nothing?" we are now led to ask, "Why music rather than noise?"

When I speak of "music," I am speaking metaphorically. It is a generalized sense of music that I mean, a bit like the ancient idea of the "music of the spheres," not limited to celestial bodies, but now including atoms and molecules as well. It is anything that manifests the magnificent orderliness of our cosmos. In order to write music (in the literal sense of the word), the composer chooses a certain number of elementary tones. He places them in a particular sequence that will unfold with time. If these tones have been chosen randomly, and if there is no relation between a tone and the ones that precede and follow it, we have "noise." If they are ordered according to a particular structure, whether that of J. S. Bach or that of the Beatles, we have music. There is an infinite number of ways to make noise, but a much more limited number of ways to make music.

A swallow soars gracefully across my window, carrying food to its nestlings. The biological sciences have shown us the extraordinary degree of arrangement and material organization and the fantastic number of perfectly synchronized chemical reactions that are hidden behind this simple event of everyday life. In our analogy, this is a manifestation of music in nature. But what are the "tones" of this music? They exist at many levels. Living things are structures made of cells, which are themselves structures of macromolecules (proteins and nucleic acids), which are structures of more modest molecules (amino acids and nucleotide bases), which are structures of atoms (carbon, nitrogen, oxygen, hydrogen, and so forth), which are structures of nucleons (protons and neutrons), which are structures of quarks. Does the structural scale stop there? No one today would dare assert that it does.

What is evident, in any case, is the hierarchy of structures. Some authors have compared it to a set of nested dolls, like the *matriochka* of painted wood that tourists bring back from Russia. But—and this is the crucial point of our discussion—we might easily imagine a situation in which these unities *never* evolved. We might imagine that, out of the almost infinite multiplicity of possibilities, only "noise" might have been born. Yet there is music. Why? This question arises at every level: Why did nucleons structure themselves into atoms? Why were atoms able to form molecules? And so on.

The first answer that comes to mind is that, in nature, there are forces. These forces create bonds and are thus responsible for the existence of all "bound systems": nuclei, atoms, molecules, cells, planets, stars, and galaxies. The organized structures that we have encountered at all levels of cosmic evolution are absolutely dependent on the presence of such bonds. We are tempted to say that "musical" unities are structured because there are forces at work between their parts. But is that really an explanation? Which comes first, force or structure? Let us first recall a bit of history, starting with gravitation. We could say that Newton first discovered the existence of a *force*: The apple fell. He then *deduced* the existence of structures bound by gravitation: the Earth–Moon system, Jupiter and its satellites, and the solar system including the Sun, the planets, and myriads of smaller bodies. Then came electricity, at the beginning of the nineteenth century. Again it was the force that was considered first: that between charged spheres. It was nearly a century before it was comprehended that atoms and molecules are systems bound by the electric force. In the case of the nuclear force, the sequence of events was reversed. The physicist Ernest Rutherford first discovered, at the centers of atoms, nuclei composed of several units (protons and neutrons). Then Enrico Fermi and his collaborators demonstrated the existence of a very powerful force that bound these units together and gave the nucleus its enormous cohesion and stability. They had discovered the nuclear force.

Which comes first, force or structure? Neither one nor the other. One could infer the existence of structures from the existence of forces, or the existence of forces from the existence of structures. One could infer both of these notions from a third (a symmetry principle, for example) (*N18*). One could also infer the third notion from the other two. All these approaches are equivalent. This is a familiar situation in physics. In the beginning there is always some basic idea that is postulated without discussion. A scaffolding is erected upon this base; we theorize, we formalize, we connect this postulate to others. But we never start from zero. We can never prove *everything*.

We could begin by saying, "There is music." We hear it, we try to understand its structure and penetrate its harmonies, but we still do not know *why* there is music rather than noise. Better to admit it at the outset: We pose as a principle that there *is* music.

What Kind of Music?

"Music rolls . . . but not from the organ."
Walt Whitman

We can classify music according to the amount of freedom that it leaves to the performer. In classical music the parts are completely written out in advance. Each note is foreordained to a particular order and an immutable tempo. In contrast, jazz sessions leave a great deal of freedom to the musician. Beginning with several more or less concise themes, the artist improvises according to his inspiration. His music reflects his audience. Each session is a unique event, never repeated, a happening.

Where does the music of nature fit within this context? Was all of the unfolding of the universe in space and time written out already in the play of interactions between particles? The flight of a swallow, Beethoven's last sonatas, or your next weekend at the beach—were they already composed in parts that quarks, electrons, and photons prepared for reading and performance 15 billion years ago?

During the first moments of the universe, no structure, no architecture, no organized aggregates exist. The extreme and omnipresent heat inexorably foils every attempt at bonding and stabilization. Everything is fluid and in motion, as at high tide. The rocks uncovered at low tide still exist at high tide, even if I cannot see them. With diving equipment, I can visit them at any time. I can make a topographic relief map. I can assure myself that, except for minor details (the disposition of the sand and pebbles), the retreat of the water does not change the mineral landscape. Is this also true of the universe as the tide of heat ebbs? Does the universe "know," while it is still bathed in the initial destructive heat, what form it will take during cooling? To borrow an image from biology, are the nuclear and atomic "niches" that will later be offered to it foretold in some Great Book?

The Search for Stability

A utopian quest.

A pile of pebbles sits at the summit of a high mountain, quite at the mercy of the elements. Storms, avalanches, earthquakes, and mountain climbers will all help nudge the pebbles downward to the plain below. The system pebbles-on-plain is more stable than the system pebbles-on-mountain-summit. We might conclude from this that the future of the pebbles is perfectly determined by their quest for stability. They will move inexorably toward the plain.

In similar fashion, the musical parts played in the universe should, according to some authors, be determined by the fact that matter constantly seeks to achieve the state with the greatest stability. As the ocean of heat ebbs throughout the universe, particles settle into "niches" made available to them by the forces of nature. They continuously seek out states of ever higher stability. One might, in fact, be tempted to describe the history of the universe as that of an ensemble of particles successively occupying all available niches in such a way as to ensure the greatest stability. In this case, the cosmic music would be predetermined, the parts would be in the style of classical music, and invention and fantasy would have no place.

A more thorough analysis reveals a quite different situation. In its most stable state, the universe would consist entirely of iron atoms. But today, after 15 billion years, less than one atom in 30 thousand is an atom of iron. Furthermore, after a rapid increase at the beginning of the life of the galaxy, this number has grown ever more slowly. At the rate things are going, iron will never have a thousandth of the abundance of hydrogen. Nuclear stability will *never* be attained. Why? Essentially because the expansion is too rapid. If primordial nucleosynthesis, instead of lasting several minutes, had stretched out over millions of years, the universe would today be made of iron, and we would not be here to speak of it. (In technical terms, we say that the universe is not in a state of nuclear equilibrium, though it was during the first few seconds. The departure from this equilibrium, which was the guarantee of maximum stability, corresponds precisely to the nuclear awakening that started the initial nucleosynthesis.)

To return to our comparison, one might say that the pebbles fall into marshes on the flanks of the mountain and never reach the plain. Under these conditions, the search for stability is no more than a gross tendency of matter. It describes only one general aspect of the unfolding of events. The evolution of the universe cannot be reduced to a quest for the most stable state. Let us imagine, in addition, that our mountain has very complicated relief, with numerous valleys on its slopes. Many of these valleys are at nearly the same altitude. According to the terminology used above, they have the same degree of "stability," intermediate between that of the summit and that of the plain. From the point of view of a descending pebble, they are equivalent.

This situation is similar to that at the gaming tables in Monte Carlo. A white ball is placed at the center of a large horizontal wheel of slightly conical profile. The players place their bets on the numbered compartments at the edge of the wheel, toward which the ball rolls. If the proprietors of the casino are honest, all the compartments are equivalent, and the quest for stability will not steer the ball to any particular place. The gradual growth of large molecular structures resembles the course of a stone that is "free" to choose its valley, or of a ball that is "free" to choose its compartment. They all have access to a multitude of equally stable states (*N19*).

In summary, for self-organizing molecular matter, the quest for stability is only a weak directional guide that is content to offer general indications. It cannot specify an itinerary among the innumerable possible paths.

Jazz

Why is there music rather than noise? We have not found the answer. We have had to accept that music exists. But, again, what is the nature of that music? Does it unfold according to a score that is fixed in advance down to the smallest details, or is it improvised as it goes along?

The story of the great steps of nuclear and chemical evolution has not shed much light on this question. We have seen the importance of forces and of bound systems throughout the growth of organization. For a moment, we thought we saw in the quest for stability a score that might govern events. This hope has not stood the test of analysis. The quest for stability, like all the laws of physics, explains the behavior of matter only in a fragmentary and episodic way.

There is, however, a partial answer to our question concerning the nature of the music. It comes from observations neither of the heavens nor of atoms, but rather of the plant and animal world. Who has not been struck by the richness of species and the variety of forms revealed in even the most modest exhibit of butterflies or seashells? On the Earth, life proliferates in all directions. It adapts to all possible locales, to all imaginable conditions and even some unimaginable ones! The bounds are set by the surface of the planet. The Earth presents a large variety of physical environments: tropical, glaciated, dessicated, moist, and aqueous. It changes over the ages. The continents move; ice ages succeed hot periods. Life must adapt, must constantly change its aim. If species disappear, other more versatile, more resistant, better performing species take their place. Nature does not invent one, but rather a hundred ways to solve each problem. I am thinking, in particular, of the countless conventions and caprices that plants and animals use to reproduce (*N20*). I am also thinking of the great migrations of animals. Life, not content with having invaded and conquered the entire planet, enjoys traveling in every sense. Tiny birds, the buntings for example, push their physical performance to the limit in crossing the oceans. They return to the region of their birth, always to the very same thicket, by guiding on anything that presents itself: Sun, stars, magnetic fields, shorelines. Their guidance system is still largely a mystery to us.

This playful and profligate aspect of nature has long been familiar to Hindu mystics. For them, nature is the creative expression of divinity, an activity that is not predetermined but analogous to a work of art. Brahma is the supreme poet. Nature (called in this sense "lila") is his means of expression. We shall tentatively conclude, in all naiveté, that the variations and almost unlimited caprices of terrestrial life are more like a jazz performance than a rigid classical score. The recent progress of molecular biology will serve to strengthen this conclusion.

The Taming of Chance

Surely God does play dice. But He ignores the losing rolls.

The ancients asked why dogs never gave birth to kittens, and why a piece of chicken that a person eats does not cause him to grow feathers.

To answer these questions, let us return to the interior of the cell and make use of an analogy (*N21*). We shall compare plants and animals to immensely complex factories, humming with activity. Each factory is made up of a large number of distinct workshops (the cells). Each workshop contains a strongbox (the nucleus), which stores the working plans (the genetic code). The plans contain detailed instructions for all the operations planned in the factory. These instructions are codified in a number of volumes (the chromosomes), of which human beings possess forty-six. Each volume contains a number of pages (the genes). Upon these pages are arrayed the letters (the nucleotide bases). Our Latin alphabet has 26 letters. The alphabet of our genes contains only four: A, C, G, and T. A sequence of these letters makes up the genetic code, just as a sequence of Latin letters makes up the civil code (*A3*). The plans never leave their strongbox. Technicians (molecules of messenger RNA) are constantly stopping by to make copies, which they then carry off to different workshops (the ribosomes). There, specialized workers blindly follow the plans and carry out the operations they are assigned. The factory functions well, but it ages and deteriorates. It must replicate itself and create new units of production. This, too, has been foreseen: The genetic code also contains the plans for the factory itself.

In this almost perfect organization, there is one weak point: the preservation of the plans. Despite all the precautions and vigilance, alterations sometimes occur. Inversions take place, letters are misplaced or even lost altogether. What are the causes of these "mutations"? We do not really know. They may be caused by cosmic rays. I spoke earlier of these space travelers that constantly bombard the surface of our planet. Their power of penetration is great. Some particles pass completely through our atmosphere and penetrate into cells, where they knock loose a number of electrons. The affected molecules fold up upon themselves, changing their messages. Their functions are disturbed. Whatever their cause, biologists agree that mutations occur without prior planning. I accept their word for this. I know from experience that to judge the credibility of an argument in some area of research, it is not sufficient merely to know something about the area; one must work actively in it. As in skilled labor, there is in research a very important thing called "professionalism." It is acquired only from years of experience. We end up having a "feeling" for things. This is what physicists call "physical intuition."

Here again chance is at work. It has accompanied us throughout our story. In the cores of stars, as in the primitive ocean, chance has served as a "matchmaker." Its role, at the elementary level, cannot but be beneficial. But now things have evolved. At the level of complexity of DNA molecules, does it make sense to place our trust in chance? Its reputation is shady. A well-organized project generally leaves nothing to chance; strict planning is usually the best guarantee of success.

This distrust is amply justified in our domain of interest. Mutations of the genetic code cause alterations in the instructions needed for the functioning of the body. Such modifications are generally harmful. They disturb the courses of these functions and provoke catastrophes of greater or lesser importance. For some biologists, these catastrophes constitute the main cause of aging. Their accumulation provokes a kind of progressive poisoning of the functions essential to life. Occasionally, however, a modification will play a beneficial role. For example, it might accelerate the rate of a chemical reaction that produces an important hormone. Individuals favored in this way will profit from this advantage (which is usually minimal) throughout their lives. Mutations that take place within the reproductive cells (in the ovaries and testicles) will sometimes be transmitted to children and grandchildren. The mutation may have the effect of improving the adaptation of individuals to their environments. They will have a better chance of reaching the age of reproduction and thus of transmitting this advantage to descendants. There would then appear a new population of individuals endowed with this mutation. It will increase faster than the population that lacks this advantage and will soon take over the whole territory.

We should pause here for a moment to express our admiration. By an extraordinary inversion, chance, generally known as an agent of disorganization and disorder, now becomes the agent of organization itself. Nature "knows" how to make biochemical structures that permit winning rolls to be retained, while losers are ignored. This is natural "selection." Einstein said, "God does not play dice." Not true. God loves to play dice, but in His casino the sympathetic croupiers ignore the losing rolls. How did this casino come into being? Is it the fruit of pure chance? Biologists do not agree on this; the debate is still going on.

Just as prehistoric man had to succeed in bridling the horse in order to make it into a powerful helper, nature, through the invention of DNA, has "bridled" chance. Jacques Monod, in his book *Chance and Necessity*, has well illustrated the importance of this step. The fundamental units of life (proteins, enzymes, nucleic acids) contain thousands of atoms. The possible combinations are almost innumerable. (They can be made with almost no change of energy. We reencounter here the freedom conferred by "degeneracy." The constraints imposed by the quest for stability are almost nonexistent.) It is a gigantic roulette wheel corresponding to billions of new configurations and billions of possible functions. One must appreciate the remarkable flexibility of this instrument and its ability to adapt (in the sense of natural selection) to varied conditions. The immense variety of animal and plant forms whispers in my ear that this idea about DNA and bridled chance must contain some element of truth.

We are now far from the rigorous conditions under which nuclei and molecules came into being. There, the energy balance severely limited the possibilities. Here, free of those constraints, an overflowing and abounding luxuriance is the rule. It is nature in its playful and generous aspect—the "lila" of the Hindus. The reader of Monod will notice, however, that my vision of events differs from his. It is a question of interpretation. From biologists I have learned the facts. These facts have been acquired by means of a scientific technology that presents every appearance of objectivity. But the interpretation of these facts involves the entire person, including one's logic, emotions, impulses, and prior experiences. Interpretation involves both obser- vations and the observer; it cannot, therefore, be wholly "objective." Each person has his own mode of interpretation that we ought to respect but not necessarily adopt. To Monod, the essential role of chance in biological evolution proves the absence of any "intentionality" in nature. On this basis he denounces as illusory the ancient alliance of mankind with the universe; he sees man as an accident in a cold and empty cosmos, a child of chance. Indeed, but a child of *bridled* chance. Let us raise our hats to nature, which has tamed chance and made of it an admirable ally.

The Anthropic Principle

"If it were not for this fact, we would not be here to speak of it." Many times in the course of this discussion we have come across apparently fortuitous events that appear to have been indispensable for the appearance of human beings. A quite extraordinary coincidence involving certain nuclear parameters made possible the birth of carbon in red giant stars. The relative populations of photons and nucleons happened to give the universe the longevity needed for the appearance of life, and at the same time permitted the formation of stars and galaxies. The list of these "miraculous coincidences" is long.

The anthropic principle has been invented, if not to explain, then at least to clarify this astonishing situation. The principle may be stated roughly as follows: "Given that an observer exists, the universe must have the properties necessary to produce that observer." Cosmology ought to take into account the existence of cosmologists. These questions would not be asked in a universe that did not have these properties.

One might reasonably object that matter has formidable adaptive abilities. On many occasions throughout our epic of life, natural en- vironments have changed in such a way as to increase their organi- zational ability. Inside stars, gravity comes to the aid of nuclear evolution and successfully carries out the project that failed during the first few seconds. The rate of formation of stars and consequently the rate of nuclear evolution itself are accelerated by the formation of heavy atoms in stellar furnaces. On our own planet, the earliest life forms transformed the atmosphere. Carbon dioxide was replaced by oxygen, a molecule eminently better suited for energy transport. On the Earth, animals have established themselves in the most inhospitable climates, making the best of the adverse circumstances. A glance at our neighboring planets shows, however, that this adaptability has its limits. The aridity of the Moon and Mercury and the high temperatures of Venus got the better of the frenetic urge toward life. Even Mars, where the conditions are not so severe, appears to be sterile.

The elaborative power of matter obviously has its limits; it does not flourish irrespective of conditions. We must nevertheless admit that the properties of the universe "allow" the advent of observers, at least in certain places. And thus we come back to the anthropic principle (*N22*).

The Adventure of the Universe

"I believe a leaf of grass is no less than the journeywork of stars."
Walt Whitman

"How futile our human preoccupations must appear to those who live in the realm of the stars." This is a remark that I have often heard. I always register a lively protest. First of all because, even if the stars are very large, their degree of organization is infinitesimal compared to that of the smallest violet in the woods. Stellar machinery is simple. It puts into play enormous energies, but its use of them is brutish. With an infinitely smaller energy supply integrated into a system of highly sophisticated biochemical cycles, the violet grows, bursts into delightful bloom, and strews afar the seeds that ensure its reproduction. My second reason is that the preoccupations of stars and of human beings are by no means independent. Throughout this book I have tried to show that humans are part of a lengthy history that involves the entire universe from the time of its birth. Our nucleons were born in the great original fire; they were assembled into nuclei in the fiery hearts of stars. These nuclei have clothed themselves with electrons to make atoms and simple molecules in interstellar space. In the primitive ocean and on the continents, the combinations continued tirelessly. At each stage, new levels of complexity have appeared. Twenty million years ago the monkeys were the most highly organized and functional beings on the Earth. Today the torch has been passed on; it has been entrusted to us.

What will be the future of this evolution? Toward what new perfections is the cosmos directing itself? What gestating plans are ripening within us? Of what are we the seed? We shall probably never know. We shall not watch the blossoming. But we have been entrusted with a mission: to assist this blossoming by every possible means, just as a pregnant woman must take care of herself. Today, this mission takes on a new dimension. Neither dinosaurs, nor monkeys, nor even humankind prior to this century had the power to destroy themselves. But today we do have the ability to interrupt the jazz concert.

"We," the reader will realize, is more than just you and me. It is the entire "adventure of the universe" that plays itself out *in* us and *through* us. Knowledge of the cosmos is much more than a luxury for cultivated souls. It is the foundation of a cosmic consciousness. It casts light on the heavy responsibilities that have fallen upon us. The most serious threat is obviously that of nuclear armaments. The arsenals of the superpowers are now more than sufficient to kill each of us 40,000 times over (*N23*). This fact carries the evocative name of "overkill power." Many times already we have passed within a hair's breadth of the holocaust. Far from being reduced, this arsenal is growing in strength and precision. From year to year, more nations join the "nuclear club."

How will all this end? We cannot exclude the worst. Might the adventure of the universe be fundamentally absurd? That adventure can be summarized in the following way. Under the influence of the forces of nature, particles join together and coordinate themselves. As the number of combining elements grows, structures increase their ability to carry out ever more complex operations. The capacity for using the physical environment, which is already present in rudimentary form at the level of animals (monkeys gather fruit with a stick), develops prodigiously at the level of humans, where it takes the name technology. Guided by science, which it itself spawns, technology has placed ever more powerful sources of energy at the disposal of humankind. As soon as they were mastered, though, fire, dynamite, and nuclear fission were used for war and destruction. The extreme instability of the situation thus created could soon provoke the end of the cycle and a return to the initial state (figure 49).

From this line of reasoning there arises a dismal image: thousands, millions of dead planets covered with the poisonous debris of their irresponsible civilizations. Is this the reason why we have not received interstellar communications? Is such fatalism justified? Is it already too late? Our only hope of survival rests on *a rising consciousness on a world scale of the extreme gravity of the present situation* (*N24*).

Can we do nothing? It is not clear. The infernal cycle of escalation will be broken once a sufficient number of people have expressed

Figure 49
The atomic bomb (Bikini) and Hiroshima, or how to kill the music. (USIS)

their *unconditional* opposition. We can no longer play at war. The music must be saved.

In the same spirit, we must, in my opinion, view with great concern the expanding nuclear power industry. The security problems have not yet been resolved, at least not to the extent that they should be. But that is not what I want to talk about here. In order to construct bombs, one needs plutonium or enriched uranium, which are direct products of the nuclear industry. The proliferation of nuclear power stations can only encourage the proliferation of nuclear arms. Each kilogram of plutonium that is produced increases the gravity of our overt insecurity. Humanity needs energy, but we must take the long view. Nuclear solutions, which are both dangerous and polluting, will not be sufficient. Solar energy alone can provide for the needs of the population of the Earth for the next 5 billion years.

III

Behind the Scenes

In a theater there is always a lot happening behind the scenes. Supporting the brilliant performances of the actors are understudies, stagehands, and many other invisible but indispensable characters.

It is now time for us to visit behind the scenes of cosmic evolution. There we shall find at work a number of entities that are both familiar and enigmatic. We do not really know what time, space, matter, energy, force, chance, and the laws of nature are. They cannot be tied up neatly in definitions. We have some fragmentary intuitions, but when we hold these up to close inspection, they lead us directly into profound mysteries.

14

Cosmic Time

"Chaque atome de silence/Est la chance d'un fruit mûr!": Every atom of silence/Is the chance for a ripe fruit. The epic of cosmic organization is structured in time. Every second, something ripens a little. Nature does its work in secret and blossoms in its own good time.

Do we know what is hidden behind this complex reality we call time? Since Einstein, the very notion of time has become rather complicated. There is no longer "one" time, but rather an infinite number of times, each with its own rhythm. The rate of passage of time is related to the speed of whatever measures it. It is also related to the amount of mass in the vicinity of the measurer. In this new perspective, what sense can we make of the "cosmic time" in which the history of the universe is written?

These are the questions that I shall address in this chapter. We may indeed rightfully speak of a "cosmic time." It has an advantage over all other times: convenience. The same story could be told in any other time whatsoever; nothing would be fundamentally different, but everything would be much more complicated.

Time, Space, and Speed

The theatrical director who stages a play must fill two different "containers." First, there is the stage, an empty space with prescribed boundaries. It is furnished with props, sets, and people—the actors. Then there is the duration of the play, a time interval that is equally empty and bounded (three hours, for example). The director places events within this time. The heroine goes on stage fifteen minutes after the beginning of the play. She dies two hours later. Time and space make up the inert and independent framework that is filled in a particular way. Their sole connection with the contents (props, events) is to contain them.

The "classical" physicist, prior to Albert Einstein, certainly would have adopted the same viewpoint and would willingly have extended it to the entire universe. The theories of Galileo, like those of Newton, were proposed in the context of fixed and absolute time. The "cultured" person would likewise have subscribed to this way of looking at things. Why? "Because it is obvious; because it is sensible." By what psychic operation is a person led to invoke "obviousness"? To say "It is sensible" is to place great confidence in the human spirit. It is to suppose that, by lonely reflection, one may arrive at a certain "truth." But can this really be done? Perhaps, if all that is involved is merely passing judgment on familiar everyday situations. The farther removed one is from such situations, the more skepticism is called for. The most important steps in the progress of physics often arise from questioning the "obvious" and the "sensible." This was the essence of Einstein's genius.

The physicist today knows that, far from being independent, time and space are in fact intimately interconnected. It is necessary to take into account the speed of the observer with respect to the object being observed.

The Runaway Carnival

Let us now travel in our thoughts to an imaginary Disneyland. In a little theater open to the sky there are rows of seats filled with spectators. Beyond is all the traditional panoply of attractions: merry-go-rounds, roller coasters, brightly lit Ferris wheels. Let us imagine that the movements of these rides have all been enormously accelerated. Each seat now moves at a speed close to that of light. From the theater, and from each of these devices in frenzied movement, people watch the play as it unfolds on the stage. Two lamps are lit. We ask the spectators when these lights appeared. Those seated in the theater are unanimous: The lamps were lit at the same time. Those on the merry-go-round and other rides in motion are of different opinions. Some claim that the lamp on the right was lit before that on the left; others claim the opposite.

Einstein arrives to play the role of Solomon. He explains that space and time are influenced by speed. Events that appear simultaneous to a stationary spectator in the theater do not seem so to a spectator carried along by the movements of one of the rides. Something that appears to be on the left for one could be on the right for another. Each has a particular perception of time and space, and everyone is right. There exists neither absolute time nor absolute space, but only a "space-time complex," the perception of which depends on the speed of the observer.

The merry-go-round at a real carnival does not travel this fast. The events described above may indeed take place, but for ordinary spectators the time differences are imperceptibly small. By the use of sufficiently precise chronometers, however, one *could* observe and measure them. Several years ago, two teams of physicists flew around the world, one traveling toward the east, the other toward the west. After their circumnavigations, they verified that, relative to very precise clocks left behind at their base, time had passed more rapidly during the eastbound trip, and more slowly on the westbound trip. The differences, measured in billionths of a second, agreed with the predictions of Einstein's theory. (The speed of the airplane heading east was added to the rotation speed of the Earth; that of the airplane heading west was subtracted.)

Einstein's Dog and Langevin's Twins

A man goes for a walk with his dog. While he strolls sedately along the road, the dog comes and goes, runs a hundred meters ahead of him, returns, runs a hundred meters behind him, all at a fast pace. The dog's long tail wags rapidly from left to right. When the walker has gone a kilometer, the dog has gone five—and the tail has gone twenty-five! That evening the dog has aged less than his master, and the tail less than the dog.

Two identical twins meet at a rocket launchpad. One stays on the ground, while the other embarks on a long trip in a space capsule. He reaches speeds close to that of light, makes a turn about Sirius, and returns. According to the clocks at the launch base, 100 years have passed. His great-grandson is there to welcome him back. The traveler has changed very little, though. According to the chronometer on his spacecraft, only three months have elapsed.

These two fables serve to illustrate the influence of speed on the passage of time. But on what ground can we assert that, if this experiment with Langevin's twins were someday to become technically feasible, we would actually obtain the result described? It is that this result follows from the theory of relativity, a theory whose predictions have been verified each time they have been tested. This unbroken chain of successes is our guarantee of the value of the theory.

Matter Retards Time and Distorts Space

By the pressure of a foot on the accelerator, the pilot of a spacecraft alters time and space. But this same result may be obtained in a different way. All we need to do is add matter. The gravitational field produced by matter slows the flow of time for those nearby relative to those far away. Time flows relatively more slowly at the bottom of a valley than on a mountain peak (the peak is a little farther from the mass of the Earth). Let us consider again the case of the two identical twins, whose lives will have exactly the same length as measured, for example, by the number of beats of their hearts. Let us imagine that the first decides to live on the seacoast, while the second pitches camp atop Mount Everest. The heart of the second twin will stop about one billionth of a second before that of the first. Similarly, time flows relatively more slowly on the surface of the Sun than on the Earth's surface. This slowing of time causes a change in the color of the light rays emitted by the solar surface. They appear to us to be a little more "red" than the same rays emitted on Earth. The difference is tiny, but it *is* measurable and *has been* measured.

How do we know that matter influences space and time? Let us return for a moment to Newton lying under his apple tree. "Why does the apple fall? Because the Earth attracts it." Starting with this idea, he invents the theory of gravitation. In this theory, time and space are unaffected by their "contents." Things happen as in the theatrical production described earlier. This theory enjoys enormous success. It accounts for the motions of the planets to very great precision. One detail, however, escapes it: It cannot account perfectly for the orbit of Mercury. The axis of the orbit of this planet is not fixed in space, but instead rotates slowly around the Sun in a way not predicted by Newton's theory.

Einstein took up Newton's problem and dug deeper into it. "*Why* does the Earth attract the apple?" He reached a new answer: "It is because the Earth 'distorts' the space in which it is embedded." He reformulated the relationship between the Earth and the Moon. He said, "The Moon runs straight ahead, like a train on its tracks. But because of the Earth, the tracks are curved. They describe an ellipse about the Earth that the Moon follows blindly." We might accuse Einstein of just playing with words. What difference is there between Newton's proposition that "The Earth attracts the Moon" and Einstein's proposition that "The Moon orbits in a space that is distorted by the Earth"? It is necessary to translate these two propositions into the language of mathematics to show that they predict slightly different orbits. Contrary to Newton's predictions, Einstein's proposition correctly predicts the behavior of Mercury. In this sense it may be said to be more "true."

Consistency with reality—the capacity to account for observations to very high precision—is the ultimate criterion of scientific truth. Quantitative considerations play a fundamental role here. In the neighborhood of the Sun the twisted rails of space change the path of light rays (*N25*). That is why we can see, at the peak of a solar eclipse, stars that the disk of the Sun would normally hide from us. Their light reaches us because it has been deflected by the mass of the Sun. This fact, first predicted by Einstein, was verified during the solar eclipse of 1919. Fulfilled predictions are the touchstone of excellence of a theory. "I don't understand. How could this be possible?" asks the person of common sense. The answer is that there is nothing to "understand." This is just the way it is. We must first state the facts and know what actually happens before we try to "understand." To deny the evidence of our eyes because it does not jibe with what we believe is the policy of an ostrich. Nature need not adapt itself to our ways of thinking. Good ideas and good theories are ones that help us negotiate the curves of nature, just as a good chauffeur negotiates the switchbacks on a mountain road.

Cosmic Time, Cosmic Space

Absolute time has vanished under Einstein's inquisitive gaze. Time, as has been verified experimentally, does not pass at the same rate for everyone. If this is the case, how can we speak of a history of the universe? What meaning can be attached to the term "the age of the universe"? It is *matter* that allows us to speak in a coherent way of a "cosmic time."

Let us imagine ourselves aboard a spaceship, traveling at 90 percent of the speed of light relative to the Earth. The universe, to our eyes, is a vast rainbow, blue in the direction of our movement, becoming progressively green and yellow to the sides, then deep red to the rear. Why? Because of the movements of the galaxies in space. We are rapidly approaching the galaxies situated ahead of us, and rapidly retreating from the ones behind us. This movement influences our perception of colors, "bluing" them as we approach and "reddening" them as we retreat. Let us reduce the speed of our ship. Progressively, the rainbow fades out. When we are back at our terrestrial base, we find ourselves in the familiar situation in which all the galaxies are retreating from us. The deep sky again becomes "red." (But not exactly!

The Earth moves at 30 kilometers per second around the Sun; the Sun orbits at 200 kilometers per second about the axis of the galaxy; the galaxy has a [poorly known] speed of several hundred kilometers per second relative to the center of the cluster or the supercluster. All these speeds are small, less than 1 percent of the speed of light. They nonetheless impart an "intrinsic" speed to our terrestrial base which makes the color of our deep sky not perfectly uniform. These tiny differences in color have, however, been detected.)

We call "cosmic time" the time of those observers for whom the deep sky is uniformly red. This time is no "better" than that of any space traveler. All times are "valid." But this one has the advantage that it applies equally to the Earth, the Sun, and the stars. It is the time experienced by *a very large number of atoms.* It is the time of most matter. It is with respect to this standard of time that we have measured the age and size of the universe.

15

Energy, Forces, and Elsewhere

Presentation of the Great Elsewhere

I would now like to present to you one of the most important characters, and one of the most discreet, in our story. We shall call it the Great Elsewhere. It is the vast expanding space between the galaxies. Without it, the organization of matter would not have been possible. Without it, the universe would never have mounted the steps of complexity, and we would not be here to speak of it.

To understand its role, we must return in a more quantitative way to the forces of nature. In the following pages, I shall present the idea of binding energy. Then, by means of imaginary experiments, I shall apply this idea to each of the forces. We shall see that no system can be formed without emitting energy, and that no energy can be emitted unless there is an "elsewhere" to receive it.

Energy as the Coin of the Realm

Behind change are things that never change.

Change is a characteristic of the world. Hot becomes tepid, bodies fall, the fire burns, and the logs are consumed. These transformations do not take place in an arbitrary way. They are interconnected by a kind of currency exchange. The money is energy. It allows physicists to interrelate the phenomena they study.

In a cannon, a charge of powder explodes. Some chemical energy (of electromagnetic origin) is transformed into kinetic energy (the shell is thrown out), and some into thermal energy (the cannon heats up). The sum of the kinetic and thermal energies is equal to the amount of chemical energy liberated.

We must have a unit of currency. At the bank, we may use the dollar or the franc. In physics, there are many units. The most useful for our purposes is the "electron volt." The name of this unit might lead us to suppose that it can only be applied to electrons, but such is not the case. In the same way that gold coins are not confined to the purchase of jewels, this unit is also useful on a broader scale. Here are a few examples. A proton traveling at 15 kilometers per second has a kinetic energy of 1 electron volt. An electron, which is much lighter than a proton, would have the same kinetic energy if it were traveling at 600 kilometers per second. The energy of photons can also be expressed in electron volts. Blue photons have about 3 electron volts; red photons about 1. The rainbow is made up of photons carrying 1 to 4 electron volts; x rays carry thousands, and gamma rays millions. At the other extreme, the fossil glow is made up of photons carrying a thousandth of an electron volt. Radio transmitters send into space photons with about a billionth of an electron volt. Relative to events in our daily lives, the electron volt is a minuscule amount of energy. One calorie is equal to 26 billion billion (2.6×10^{19}) electron volts, while one liter of gasoline carries a billion billion billion (10^{27}) electron volts.

The expression, "the law of conservation of energy," means that the amount of energy *before* any event is exactly equal to the amount *after* the event. Is this law absolute? During the 1930s, physicists discovered the existence of the neutron, an unstable particle (with a half life of about 15 minutes) that decays into a proton and an electron (*A3*). The energy balance revealed less energy after the disintegration than before it. Nonetheless, confident in the value of the law, the physicist Wolfgang Pauli proposed the existence of a new particle, quite invisible, that is emitted at the moment of disintegration. This particle, called a "neutrino" (little neutron), must, by definition, carry exactly the amount of energy needed to balance the energy accounts for the reaction. Several years later the neutrino was detected in the laboratory. It has since taken on great importance in physics and in cosmology (*A2*). This event is significant, for it shows that the idea of energy is fruitful and well adapted to reality.

The law of energy conservation is not, however, "absolute." Its requirements are related to the duration of the phenomenon being observed. The size of the allowable "violation" of the law is inversely proportional to that duration. (More precisely, we must point out that the "violations" originate from the fact that, if the duration is very short, the energy is poorly defined.) Everything happens *as if* the energy were not rigorously conserved. These differences play a fun-

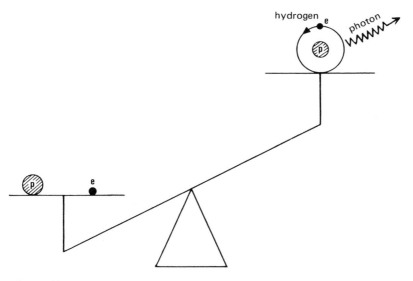

Figure 50
The electromagnetic bond. The separated proton and electron are heavier than
the bound system formed by the two particles (a hydrogen atom). The mass difference is emitted in the form of energy (an ultraviolet photon) at the moment of
combination. This mass difference, about one part in 100 million, is characteristic
of the electromagnetic force at the level of atomic and molecular structures.

damental role in the behavior of individual particles. At the level of
our everyday reality, which involves myriads of particles, they average
out and become practically negligible. The law of conservation of
energy then takes on considerable precision.

Electromagnetic Bonding

The burden of freedom: Bonding makes things lighter.

Using an extremely precise balance, we shall set to work executing
several instructive experiments. First, we shall weigh an electron and
a proton separately and add their masses together. Then, we shall
put the electron into orbit about the proton to make a hydrogen atom,
and put the atom on the balance. Surprise! *The atom is lighter than the
sum of its parts.* The difference is tiny, only one part in a hundred
million, but it is real, and that is what matters here (figure 50).

To understand this result, we must introduce a new sort of energy:
the energy of mass. We may transform mass into thermal energy or
into kinetic energy. Conversely, we may also transform light energy,

for example, into mass. (This is what is expressed in Einstein's famous equation $E = mc^2$. One gram of matter yields 6×10^{32} electron volts.) This is done every day in nuclear physics laboratories.

Let us return to our experiment. We have joined a proton and an electron to make a hydrogen atom. Under the influence of the electromagnetic force, the electron goes into orbit. This event is accompanied by the emission of a photon of ultraviolet light. The energy of the photon corresponds exactly to the mass difference between the hydrogen atom and the two free particles. The mass of the proton is equivalent to 938 million electron volts, and the mass of the electron is 511 thousand electron volts. The difference between the sum of the masses of the proton and the electron and the mass of the hydrogen atom is 13.6 electron volts, or about one hundred-millionth of the sum of the masses of the proton and the electron when separated. That is the energy of the emitted photon. In other words, at the moment of "capture," a fraction of the mass is transformed into the energy of the photon. The photon *leaves* the atom, carrying *away* an energy equivalent to this "mass defect." I have emphasized the words "leaves" and "away" because they are very important. We shall speak of them again, and we shall discover in them a truly astonishing dimension.

Let us now aim a beam of ultraviolet light at a hydrogen atom. If the photons have the necessary energy, one of them may be absorbed by the system. The electron will then be severed from the proton and set free in space. We say that the atom of hydrogen has been dissociated. The particles have resumed their original masses and their freedom. All atoms and all molecules, however complex, behave similarly. Dissociated, separated into their components, they are always heavier than they are in the bound state. In combining under the influence of the electromagnetic force, they emit the energy corresponding to the missing mass. This energy is not necessarily transformed into light. It may also be transformed into mechanical, electrical, or some other form of energy.

A discharged battery is lighter than a charged battery. The missing mass is transformed into electrical energy, then into light. When gasoline burns in air, chemical reactions produce new molecules. The total mass of these new molecules is very slightly less than the combined mass of the gasoline plus the oxygen. The difference has been transformed into heat. By means of pistons and connecting rods, I convert that heat into mechanical movement. My car runs. In parallel, I transform some very unstable molecules (gasoline) into much more stable molecules (carbon dioxide and water vapor). It is the energy gained at the time of creation of these stable molecules that I use to get my car to move. You eat a steak seasoned with pepper and herbs from Provence. The chemical reactions that ensue are extraordinarily numerous and complex. If, through a detailed study, you could add up the masses of all the molecules after digestion, you would find once again that their sum is smaller than the total mass of the initial reactants. The mass lost, after having charmed your taste buds, warms you and releases the energy that you need to turn the pages of this book. In heating our block of iron, we have carried out the reverse experiment. The iron atoms are bound in a crystalline lattice. To free them, we must use energy. Heat (thermal energy) will be transformed into mass energy to overcome the deficit. The ensemble of iron atoms is heavier than the block of iron.

All chemical reactions involve a mass change of well-defined size. The fraction of mass gained or lost varies from about one part per trillion to about one part per 100 million. Chemical reactions take place by the exchange of electric charges. This range of mass fractions is characteristic of electromagnetic energies in the realm of atoms and molecules. (Some electromagnetic phenomena necessarily fall outside this definition, but they hardly ever occur in the course of the events that are of interest to us here.)

Nuclear Bonding

On the left pan of our balance we now place a free proton and a free neutron. On the right pan we place a "deuteron"—a nucleus composed of a neutron and a proton in orbit about each other (this is the nucleus of the heavy hydrogen atom). The particles are extremely close to each other. Their mean distance is about 20,000 times smaller than the distance between the proton and the electron in a hydrogen atom (figure 51). Once again the bound system (the deuteron) is lighter than its components. But this time the difference is about one part per thousand, some 100 thousand times larger, relatively, than the case of the hydrogen atom (where the difference was one part in 100 million). These numbers clearly illustrate the strength of the nuclear bond. They also explain the economic interest in nuclear energy and the efforts that have been expended to control it. One ton of uranium yields as much heat as the hundreds of thousands of tons of oil filling the holds of a huge tanker.

As the proton and neutron combine, an energetic photon is emitted, which carries away the mass defect. Conversely, by bombarding a deuteron with an appropriate photon, we can dissociate it and liberate

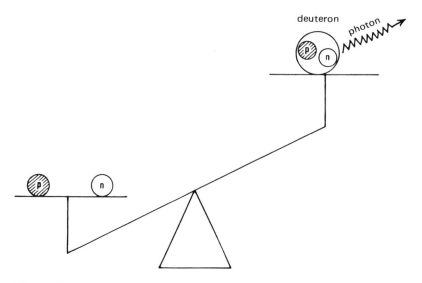

Figure 51

A nuclear bond. The proton and the neutron, in isolation, are heavier than the bound system formed by the two particles (deuteron). The mass difference is emitted in the form of energy (a gamma ray) at the moment of combination. This mass difference, about one part per thousand, is typical of nuclear forces.

the two nucleons that constitute it. That proton must be a gamma ray with an energy of 2.2 million electron volts. Since the masses of the proton and the neutron are each in the neighborhood of a billion electron volts, we find here the fraction of one part per thousand noted above.

The nuclear realm thus displays the same behavior that we found in electromagnetism. The Sun derives its energy from nuclear reactions—from a fusion process in which four hydrogen nuclei combine to create one helium nucleus (*A4*). The sum of the masses of four hydrogen nuclei is nearly 1 percent larger than the mass of the helium nucleus. (The sum of four hydrogens is equivalent to an energy of about 4 billion electron volts. The difference between the mass of the hydrogens and the mass of the helium nucleus is 26 million electron volts, and thus close to 1 percent of the total.) At the moment of fusion, the bonding energy is transformed into gamma rays. After being absorbed and reemitted countless times by the material of the Sun, this radiation arrives at the surface in the form of yellow light: sunshine.

Three helium nuclei are fused into a carbon nucleus in the hearts of red giant stars (*A4*). This time the mass surplus released is about one thousandth of the initial mass. For very heavy nuclei, the reverse phenomenon occurs. Instead of fusing, these nuclei fission, because they are more massive than the sum of their components. Uranium, for example, releases a thousandth of its mass when it explodes. In a reactor, this energy heats water, which then drives the turbines. Iron has the most stable nucleus. A temperature of billions of degrees is needed to decompose it into nucleons.

Quark Bonding

Contemporary physics points toward a similar scheme at an even more fundamental level (*A3*). Nucleons are themselves bound systems, made up of three quarks. The binding energies are even higher than those in nuclear systems. They are in fact comparable to the masses of the nucleons they constitute. This phenomenon is illuminated only within the context of some very complex physics that I will not try to describe. In any case, we are here at the forefront of research, and nothing beyond this point is firmly established.

Gravitational Bonding

Would I now astonish you by telling you that the mass of the Earth–Moon system is less than that of the Earth and the Moon taken separately (figure 52)? The difference is a billion tons! That seems immense but in fact represents no more than one part in 30 trillion (3×10^{-14}) of the combined masses of the Earth and Moon. This is a much smaller fraction than in any of the other forms of bonding we have encountered so far. Even in the electromagnetic realm, the fraction of missing mass is at least 100 to 1000 times larger.

Consider now two black holes circling about each other in a very close orbit (*A6*). In this case, the mass defect will be in the vicinity of 100 percent—much greater than the mass fractions characteristic of nuclear and electromagnetic forces. Here we touch upon a very important point. While the mass fractions for the nuclear and electromagnetic forces appear to fall into well-defined ranges (from one part per thousand to 1 percent for the nuclear force, and from one part per trillion to one part per hundred million for the electromagnetic force), for gravitation this fraction can vary enormously. It is this fact that permits gravity to play such a fundamental role in the growth of complexity. We shall illustrate this on the following pages.

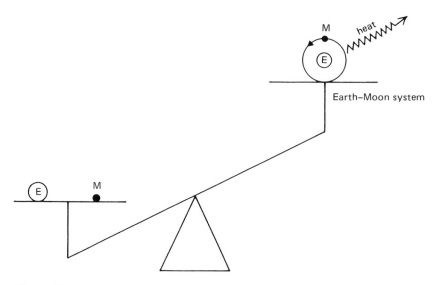

Figure 52
A gravitational bond. The Earth and the Moon, in isolation, are heavier than the bound Earth–Moon system. The mass difference was emitted as heat at the time of formation of the solar system.

The Play of Heat and Gravity

These two accomplices enjoy leaping from one realm of force to another.

Our Sun is made up of about 10^{57} particles (1 with 57 zeros). These particles (nuclei and electrons) are held together by the force of gravity. Each particle is attracted by all the others, and in turn attracts all of them. The difference between the mass of the Sun and the sum of the masses of all its particles is about equal to the mass of the Earth. This is about one part per million of the mass of the Sun.

The Sun is hot. It contains a vast amount of thermal energy. The temperature at its center rises to 16 million degrees. Why? To understand, let us go back to the time of its birth. The particles that will make up the solar matter are now dispersed within a vast interstellar cloud. These particles attract and approach each other. The entire cloud slowly contracts. It is then confined in a smaller volume (figure 53).

We can do the bookkeeping for this operation. We can show that the binding energy released by this contraction is transformed into both heat and light emission. This heat is what *remains* within the boundaries of the cloud; it is the thermal energy that agitates the

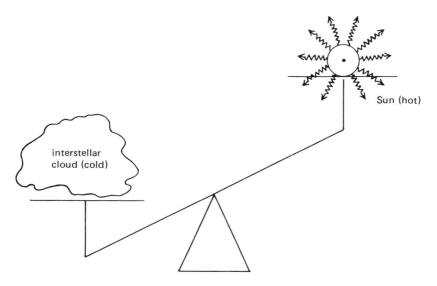

Figure 53
The mass of the Sun. The Sun is lighter than the interstellar cloud from which it inherited its atoms. The energy gained is partly transformed into radiation and partly stored as internal heat.

particles and drives them to ever higher temperatures. The light emission, on the other hand, is what *escapes*, passing through the surface of the cloud and going *elsewhere*. At the birth of the Sun, it consisted of infrared radiation; today it consists of visible light (*A1*). As the cloud contracts, it emits more and more light, while its internal temperature continually increases. When the temperature of the cloud reaches several hundred degrees, electromagnetic forces dominate the scene. When it reaches several million degrees, nuclear forces come into play.

Gravity serves as a sort of "elevator" between these realms of the different forces. According to its state of contraction (or expansion), a mass of matter will establish in its interior a temperature level that favors different kinds of interactions. It is like an oven in which one can regulate the temperature to roast a chicken or to bake a cake. It is gravity that governs the oven of cosmic cookery.

Elsewhere: An Essential Condition of Bonding

Bound systems (nucleons made of quarks, nuclei made of nucleons, molecules made of atoms, galaxies made of stars) must, at the time of their formation, emit surplus energy. This energy then goes "elsewhere." No structure, no organization of matter, can come into being unless there is an "elsewhere" to absorb the radiation that escapes.

Where is this "elsewhere"? In a stationary universe (without expansion), there would be no elsewhere. All the radiation emitted in the creation of one system would someday be absorbed, leading to the dissociation of another system. At each moment, the number of atoms formed would be equal to the number of atoms destroyed. The net growth of structure on the scale of the universe would be zero. In other words, the "elsewhere" of one system would be the "here" of another. No rocks would emerge definitively from the ocean of heat.

In a very real sense, it is the expansion that produces the "elsewhere." That is, it creates the possibility that most of the radiation will never again be absorbed. The photons given off reach intergalactic space. There they travel through a universe that is ever more dilute, ever emptier. As time passes, their chances of being reabsorbed grow continuously smaller. Time plays in their favor. The universe is transparent toward the future. Its transparency is the guarantee of the longevity of structures.

Kepler asked why the night sky is dark. He reasoned that in an infinite universe the entire celestial vault ought to have the brightness of the surface of the Sun. We have seen that the darkness of the night sky is related to the youth of the universe. Now we shall go farther. Without expansion, stars would not be able to contract and form. No nuclei, no atoms would have been assembled. The existence of helium nuclei, like the darkness of the night, is due to the universal expansion. (I do not attempt to offer a proof here, only coherence arguments. All we can say, strictly speaking, is that the helium nuclei are not in equilibrium with the gamma rays emitted at the moment of their formation.)

Elsewhere: An Essential Condition of Organization

On the preceding pages we have seen how the forces of nature create bonds between "elementary particles" of matter (*N26*). Particles and bonds, like bricks and mortar, are the essential elements of all architectures. Thanks to them, matter becomes ordered and organizes itself into complex systems.

In order to describe this ordering of nature, it is useful to introduce the ideas of "information" and "entropy." The cathedral at Chartres is built of a great number of stones, put in place by specialists according to well-defined instructions. We might say that the cathedral contains more "information" than the initial pile of stones. This increase of information is proportional to the number of instructions given by the architect. Chartres thus contains more information than a simple country church. Entropy is, in a specific sense, the opposite of information. It is a measure of the disorder in a given object. The cathedral, once finished, has a tendency to deteriorate. Erosion by wind and rain, severe winters, urban pollution, and Earth tremors cannot do it any good. Without maintenance, its entropy will increase. It will some day revert to the pile of stones out of which it was built.

Several ice cubes float in a glass of warm whiskey. The ice cubes melt, and the liquid cools off. The first state contains two phases: liquid, with solids at its surface. The second state contains only one: liquid throughout. The first state is more ordered. It contains more information than the second; I must make more statements in order to describe it. The entropy of the second state is larger than that of the first. Our daily experience teaches us that, when things are left to themselves, they tend to become "disorderly." Cathedrals fall into ruin, and the ice melts in warm water. There is a principle (the second law of thermodynamics) that expresses this conventional wisdom. It states that, in the course of evolution of events, entropy increases (or at least does not decrease).

The history of the universe seems to manifest the opposite principle. We have witnessed the inexorable progress of matter toward the highest summits of organization, with nucleons, nuclei, atoms, molecules, cells, and organisms as steps along the way. In this sense, the appearance of life seems to run counter to the natural tendency of events.

A familiar example will help us to resolve this paradox. When winter comes, lakes freeze. But ice is a much more orderly structure than liquid water. The molecules in ice are arranged according to a precisely determined geometry. In the liquid, the molecules are free to move randomly, and they constantly change their positions. The entropy of liquid water is much higher than that of ice. How can the lake acquire the necessary orderliness for the molecules in it to form crystalline ice? The key to the paradox is that, in freezing, the lake gives off a large amount of heat. This heat, in the form of infrared radiation, carries its own entropy as it leaves the Earth and heads for intergalactic space. We can calculate the amount of entropy that has been carried off by the infrared photons and show that the sum of the entropies of the ice and the radiation is *greater* than that of the lake before it froze. Thanks to the electromagnetic force that binds the molecules to each other, the lake has acquired some *local* order. But at the same time it has given off radiation of high entropy. Its increase of organization occurs at the expense of an increase in the disorder of the universe.

The same is true of the helium nuclei born during the first few seconds of the universe. They are produced by the joining together of four nucleons in the same place. As we have seen, the mass of a helium-4 nucleus is about 1 percent less than the sum of the masses of the four nucleons taken separately. This mass defect is transformed into energy that is "emitted" in the form of light radiation. "Emitted" means rejected to some place far from where it is produced. That place is elsewhere. In fact, it is added to the universal radiation field. It warms the universe very slightly, or, more exactly, it retards its cooling, albeit imperceptibly. The helium nucleus has an internal structure: The nucleons confined within it are bound in stable orbits with well-defined properties. The nucleus is more "organized" than the four independent nucleons from which it is formed. This organization appears spontaneously in the universal soup. Besides carrying away the mass surplus, the photons emitted during nuclear fusion also carry off entropy. This is the entropy that helium must disburse if it is to be allowed to structure itself without violating the principle of increasing universal disorder. These helium nuclei are tiny islands of order in an ocean that is now more disordered than ever before.

This pattern is quite general. We find it at every stage of organization. In the great universal disorder, islands of orderliness spontaneously appear. The acquisition of this organization is costly. It is accompanied by a further increase in the cosmic entropy. This entropy is "deposited" in the Great Extragalactic Elsewhere. Thus, over the course of the ages, an ever smaller fraction of the mass of the cosmos, disseminated on a multitude of favored islands, establishes an ever higher level of organization, to the detriment of an ever more disorderly universe. In the preceding section we saw that, without expansion, no stable bond could form. The elsewhere is required by the principle of conservation of energy (also called the first law of thermodynamics). We have now gone even farther: Without the elsewhere, no organization could be acquired by matter. The elsewhere is an indispensable precondition for the formation of islands of organized matter, both by reason of the conservation of energy and by reason of its inexorable degradation (the second law of thermodynamics).

We have arrived at the end of a long train of thought. To the question "Why is the night sky dark?" posed at the beginning of the book, we now see that the phenomenon of galactic recession and universal expansion can provide an answer. In lapidary terms we may add here, "If the night sky were not dark, there would be no one here to pay attention to it."

16

Chance

Throughout our epic, chance has played a role of prime importance. Among our cast of characters, however, it is certainly the most mysterious and the most ambiguous. Is there really such a thing? Or is it simply a cloak for our ignorance? Before addressing this question, we must carefully distinguish between two kinds of chance. There is that of our daily lives ("Imagine, running into you here!"), and that, more fundamental in nature, in the private lives of atoms. In an imaginary anecdote, we shall find both at work in the Tower of London. Modern physics teaches us that both kinds of chance rightfully exist, and that knowledge cannot eliminate them. We shall see how they coexist peacefully with the laws of nature and how, taken together, they form the canvas upon which the tapestry of complexity is embroidered.

Causality and Chance

Why? Because. Effect and cause. This pair has been very fruitful. Since ancient Greece (and perhaps even before), they have served as the handmaidens of philosophy and science. We observe phenomena, and we assume that they must have had a cause. We attempt to find that cause.

The idea of causality is forced on us by its richness and usefulness. It proves itself in practice. But exactly what does it correspond to? What does it tell us about the deeper workings of nature? It teaches us that reality is not total chaos. Everything is not left to chance. There is a certain "determinism" to things. How far does this determinism reach? Each year the progress of science reveals to us new causal relationships between events that had seemed to be unrelated. Will chance be completely eliminated someday? Will the universe be revealed to be completely determined, down to the smallest details?

Physicists at the beginning of the nineteenth century would have answered yes to these questions, if one is to believe Laplace's celebrated statement: "We can now envisage the present state of the universe as an effect of its earlier state, and as a cause of that which will follow. An intelligence that, at a given instant, knew all the forces that animate nature as well as the circumstances of all the entities that make it up, and that was also vast enough to subject these data to analysis, could embrace in the same formula the movements of the largest bodies in the universe and those of the lightest atom: Nothing would be uncertain for it, and the future like the past would appear before its eyes. The human spirit offers, in the perfection that it has given to astronomy, a crude sketch of this intelligence. Its discoveries in mechanics and in geometry, joined to that of universal gravitation, have taken it to the threshold of understanding in the same analytic terms the present and future states of the system of the world" (*N27*). This dream of Laplace, we now know, will never be realized (*N28*). And so much the better! How boring and dull the world would be without fantasy. At two distinct levels, reality offers resistance to Laplace's analytic intentions. Let us study them in turn.

The Chance of Insurance Agents

You would like to insure your barn against fire. "Here's what it will cost you," says your agent. By studying the fire statistics for the county, the company's actuaries can forecast quite accurately the number of barns that will burn down in the years to come. Making the assumption that fire strikes "by chance" (randomly), they calculate the probability that it will happen to you. But the actuaries of course cannot foresee *which* barns will burn.

Chance here seems to play the role of an alibi for the company's ignorance or, more exactly, for its lack of interest in the individual barns it insures. The company could, for example, consult an atmospheric physicist. It would learn that lightning is made up of a multitude of electric charges that have accumulated in a cloud. Each of the charges has its own individual history that involves a complex interplay of physical forces. By tracing the branching courses of cause and effect, one might, in principle, be able to predict where the lightning will strike. In practice, however, this task is impossible. In order to attempt it properly, we would need to have extremely precise information not only on the positions and velocities of all the electric charges involved in the lightning, but also on all the external forces that act upon these charges. The company might attempt to learn which barn will burn, but the project would oblige it to hire an infinite number of physicists and data processors. It would be much better for the company's budget simply to remain in a state of blissful ignorance.

The formulas imagined by Laplace to describe the entire universe run into a similar difficulty. They would have no chance of predicting the future correctly unless they incorporated an infinite number of infinitely precise measurements. The least deviation from this rule would inevitably result in a progressive deterioration of predictive power. Further, this necessity of infinitely precise measurements is not only unrealizable in practice but is also in conflict with physics itself. We shall see why in the following pages.

Chance and the Private Lives of Atoms

> The physics of atoms (which we call "quantum mechanics") tells us that chance is inscribed at an extremely deep level in nature. At that level, chance has every right to exist and cannot, in practice or in principle, be dislodged by knowledge.

Uranium-235 is an unstable nucleus with a half life of a billion years. What is the cause of this instability? Physics tells us that the nucleus splits because it carries too much electric charge—92 protons confined in an infinitesimal volume. These particles repel each other violently. This repulsion "tends to" blow the nucleus apart. What is the meaning of "tends to"? This is the issue to which we now turn.

Let us place 1000 uranium-235 nuclei on a dish and watch them. After a billion years, 500 remain; after 2 billion years, 250 remain; and so on. This statistical effect reminds us of the actuarial world. The difference is that, according to modern physics, one cannot know in advance *which* of the atoms will disintegrate. At first, they are all identical. Despite their "heavily charged" heredity, none of them has a personal history that would help us to predict its individual future. We touch here on the very limits of causality (in the traditional sense of the term as determinism). Causality describes only a part of reality. It permits me to know that, because of the high electric charge, the nuclei "tend to" fission. Applied to uranium-235, it tells me that, on average, one of the thousand nuclei will explode during the first million years. But I would not be astonished if this first explosion took place after 5 minutes, or after 10 million years. This aspect of the process lies entirely outside the realm of traditional causality.

It is normal to suppose that the theory is merely incomplete, that what we are taking as an aspect of reality is in fact only a proof of our ignorance. But on looking closer, we find that this "indeterminism" is not simply an incidental element of the theory that might eventually be eliminated. It is, rather, one of the pillars of the theory. Like Dante at the gates of hell, the physicist sees, written on the lintel of the temple of quantum mechanics, the words: "Abandon all hope of knowing the individual histories of atoms."

One might wish to reject the theory because of its arrogance. Perhaps another, less exacting theory could be found that would permit a restoration of the kingdom of absolute causality. The problem is that quantum mechanics has wrought many marvels. Its phenomenal success has justified its entry into the pantheon of science. It is, without question, a "good" theory. Many physicists have sought to replace it with some equally satisfactory but less exacting theory. A waste of time. To this day, it has no rival. Furthermore, it demonstrated, some years ago, the justification for its arrogance (*N29*). Theories that aspire to reestablish the individual histories of atoms are *in contradiction with experimental results.* (Some other theories do retain close agreement with experiment, but only at the cost of added hypotheses that are equivalent to those of quantum mechanics.)

Einstein, in particular, never accepted this state of affairs. It did not accord with his vision of the intelligibility of the world. Maintaining that "God does not play dice," he spent many years in vain attempts to eliminate this element of indeterminacy.

How does causality come to coexist with this element of chance in the lives of individual atoms? It waters down its wine. Everything happens as if the laws were no longer absolute; they tolerate infractions. I take as an example the behavior of electric charges of the same sign. The classical law requires that they repel each other. Let us fix a positive charge and fire a second positive charge at it. The law says that the second ought to approach to within a particular distance of the first. Then, under the effect of the repulsion, the second ought to stop and then start to retreat. This is in fact what happens *usually, but not always.* Sometimes the second charge continues on by, completely ignoring the presence of its sister. Other times, the closest distance achieved will not be that predicted by the law. Quantum mechanics allows all of those possibilities, and assigns a probability to each, but goes no further than that. It refuses to answer questions that lie outside its domain.

What about the Earth in its orbit? Does it run some risk of suffering from this tolerance? What if the Earth should one day decide to pass through the Sun instead of orbiting around it? This event is *not impossible!* There is a certain probability that it might happen. We can calculate it, and, fortunately, it is extraordinarily small. But it is not zero. This example illustrates a crucial point. Tolerance in the laws matters mostly at the level of individual atoms. It rapidly diminishes in importance as we study ever more numerous assemblages of atoms. At our own scale, there is no practical effect. This is why it went undetected for so long.

The Diamond in the Tower of London

The tourist who sails down the Thames in a boat will see at the water's edge the sinister Tower of London. It served for a long time as a prison. Today, it serves as the repository of the Crown Jewels. They are enclosed in thick cases of glass.

Among the many precious stones, there is one particularly magnificent diamond, as large as a hen's egg. Well protected by the glass wall, it seems completely out of reach. With a little patience, however, it could be in your pocket. Even better, it could get there in two different ways. First, by the movement of the air. In the case, the molecules dash about in all directions and strike the diamond with almost exactly equal pressure on all its facets. It could happen that, at some moment, quite by chance, all the colliding molecules might arrive on the same side. Lifted by this powerful current of air, the diamond would rise into the air, break the glass, and land in your hand. This is the chance of insurance agents. The second way to get the diamond appeals to the indeterminacy of the laws of physics. No object is assigned definitively and absolutely to a particular location. Usually, of course, we find things where we put them; but sometimes they end up somewhere else. This also is a matter of probability. There is a certain probability that the diamond placed in the case will be found in your pocket, but this time without having broken the pane of glass.

In both cases the chances are very small. I do not believe that the royal family is insured against them. But the chances are not zero. That is important.

To Observe Is To Perturb

How are things when we do not look at them?

Whoever has seen a busload of tourists disembarking in a Central American Indian village will have no difficulty grasping the justness of this title. With a little more discretion the perturbation could be reduced, but only up to a point.

This difficulty of observing without influencing also exists at the level of objects. In order to catch speeders, the police set up radar posts. Their instruments project toward passing cars packets of photons with well-determined wavelengths. The photons are reflected off the cars, and return in echo to the post. The wavelength of the reflected photons will be slightly different from that of the original photons. This difference permits the policeman to determine your speed and to know whether you are in violation of the law. What most policemen undoubtedly ignore is that, in reflecting off of your car, his radar beam has changed the car's speed! The waves carry energy. In bouncing off your trunk, they have given you some additional speed. The difference is, of course, tiny, and you will not feel the impact of the photons. You would have a hard time attributing your excessive speed to this perturbation.

If we observe atoms rather than automobiles, the problem of perturbation can no longer be so easily set aside. Here is an atom whose future I would like to predict. To do this, I must know with the greatest precision *where it is* and *where it is going*. In order not to disturb it, I shall choose to observe it with radiation of very low energy. According to theory, the lower the energy, the longer the wavelength. But here things begin to deteriorate, for it is impossible to localize an atom with high precision by using a long wavelength. For example, if I choose to irradiate the object with radio waves, I will end up with a margin of uncertainty of several tens of meters in its position. Here we reencounter the situation of the diamond in the Tower of London. The location of objects is uncertain because of the wavelike character that nature imposes on them. To measure the location of an atom very precisely, I am forced to use radiation with a very short wavelength, for example, an x ray or a gamma ray. Because of the great energy of such waves, their impact will be violent, and their perturbation will be important. In fact, whatever wavelengths are used, it is impossible to measure simultaneously and with absolute precision the speed and position of a particle. Whatever precision I gain in one measurement, I lose in the other. At best, I shall arrive at a compromise that renders the whole description approximate.

This element of "fuzziness" that nature preserves at the atomic level will affect the possibility of knowing the future. It will degrade the description of tomorrow, a description that we must have in order to speak of the day after tomorrow. The farther we look into the future. the more the contours blur. To envision the future of atoms, nature has given us myopic eyes. Recall, however, that these effects are no longer in play at the level of large masses of matter. The indeterminacy of the atomic laws does not prevent us from foreseeing the future of stars or galaxies. The behavior of individuals averages out. It is not tomorrow that the Earth will spontaneously change its orbit.

The "Laws" of Physics and Their Limits

> Explanations of even the simplest of events involve the entire history of the universe.

The Moon orbits the Earth. The Earth and the other planets orbit the Sun. All the bodies of the solar system obey the law of universal gravitation. Everything here exudes orderliness, harmony, stability, and eternity. There is apparently no place for the historical or the accidental.

Yet . . . Newton's laws require that the orbit of the Earth must be an ellipse, no more than that. They do not require that the direction of the orbit go one way or the other. They do not require that the plane of the orbit be located in any particular plane. What is it that governs the direction of movement of the Earth about the Sun? Why are the orbits of the planets and satellites nearly all in the same plane? These matters fall outside the scope of Newton's laws. We must look elsewhere.

I have taken the particular example of the solar system to illustrate a very general situation in physics. The laws of physics in fact explain only a part of reality. They describe *how* events occur *if* certain conditions are realized. They have no control over these conditions, called "initial" or "limiting" conditions by physicists.

Let us return to our example of the Earth. To find a detailed explanation of its motion, we must go back in time. We return first to the birth of the Earth. Its motion today preserves a memory of its "launching" (like the launching of a satellite). It inherited its course and its orientation from the gaseous nebula in which it was born. That nebula, whose form was a flattened disk, rotated about its own axis. It transmitted this rotation to all the bodies that formed in its midst: Sun, planets, satellites, asteroids. That is why the planes of their orbits coincide so closely; they were shaped long ago in the nebular disk. (This also explains why, as seen from Earth, they stay within the band of the zodiac.) This nebula itself obeyed a set of physical laws. The situation here, however, is much more complex. We have very poor understanding of the factors that give interstellar clouds their orientation and rotation. There is the general rotation of the galaxy, of course, but there are also local turbulences and other effects, such as the powerful magnetic ties that bind the nebulae like pearls on a necklace.

These factors might, in principle, be understood and appreciated by constructing a history of all the elements of our galaxy and all the interactions they have undergone. In practice, this task would be much less tractable than that of our insurer of barns. But let us suppose that, despite everything, we have achieved this goal. We are not yet done with our exertions, for we must now face the redoubtable problem of the origin of the galaxies, their turbulence, and their magnetic field. The only thing cosmologists dare assert is that these phenomena apparently grew out of properties of the matter that existed at the time of appearance of the galaxies. A caricature of the situation might go: Things are what they are because they were what they were. To explain so commonplace a fact as the rotation of the Earth, we must go back to the origin of the universe, into a past in which every clue we have becomes lost in the "night of time" (obviously this expression was poorly chosen; those times were bathed in the blazing fire of the original radiation).

In summary, to understand any fact or event, it is necessary to know simultaneously all the relevant laws of physics and also the bounds within which they work. These bounds imply the play of other laws within other bounds. Step by step, these bounds and laws bring in the entire universe, in time, in space. It is within this framework that chance plays its game.

Chance, an Essential Element of Cosmic Fertility

Throughout our epic, we have seen chance at work. Nuclei wander about in the fiery hearts of the stars. A collision takes place, and a heavier nucleus forms. Two molecules come into contact in the primitive ocean. They combine and give birth to a more complex system. Inside a cell, a cosmic ray causes a mutation. A protein acquires new properties.

Surely each of these particles already had the ability to combine or to be transformed. But a fortuitous event was required for that possibility to be realized. The organization of the universe demands that matter abandon itself to the games of chance.

17

Three Enigmas

We shall end by evoking three enigmatic facts that are rich in meaning. They appear in three wholly different domains. The first—the Foucault pendulum—is on our own scale. The second concerns the observation of the most distant objects in the universe. The third pertains to the world of atoms, involving an experiment carried out in a nuclear physics laboratory.

These facts share the common feature that we do not have a complete and satisfactory explanation for any of them. We vaguely sense that they might lead us to new perspectives on the nature of things. They have already allowed us to glimpse unexpected relationships among the mysterious characters in our story.

The Foucault Pendulum and Mach's Principle

> The entire universe makes its presence felt in the great hall of the Pantheon. It steers the pendulum suspended from its dome.

It is a quite ordinary pendulum, except that the cord is very long and the bob very heavy. Once set in motion by the attendant, it continues to swing for a number of hours. On the ground is spread a little hill of sand, in the form of a ring about the midpoint of the pendulum's course. A metallic point attached to the bottom of the pendulum bob cuts a groove in the ring of sand at both ends of each swing.

Let me remind you of the astonishing behavior of the pendulum. In the course of several hours, the plane in which it moves—the oscillatory plane—turns about a vertical axis. The attendant affirms that, when launched (for example) into an east–west plane, the pendulum progressively reorients itself toward the north–south plane. It continues past that point, returning eventually to its original plane.

The little hill of sand, brushed aside by the sharp point, bears testimony to this movement. Why does the plane of the pendulum move? What is the force that leads it to change its oscillatory plane? One is tempted to answer that it is the Earth that turns, not the plane of oscillation. That is, the plane remains fixed; it merely seems to turn because of the rotation of the Earth. This does not solve the problem. There is no absolute motion. Something is said to turn relative to something else that, by definition, does not turn. Which is it that turns, the Earth or the oscillatory plane? And with respect to what does it turn?

Let us repeat this experiment in a fictional setting. We imagine the surface of the Earth to be covered by a perfectly opaque cloud layer (like the surface of Venus). No one knows about the Sun. Humankind has nevertheless appeared, science has developed, and a new Foucault is busy with his pendulum. To simplify the discussion, I shall assume that he has set up his pendulum at the North Pole. The advantage of this is that the oscillatory plane makes a complete turn about the vertical axis in one day. At Paris, New York, or Tokyo, because of a complicated latitude effect, it makes only a fraction of a turn. Our new Foucault does not know that the Earth "turns." He asks why the plane of his pendulum rotates, and no one can answer him. He also asks why the period of rotation is 24 hours, rather than 36 or 71.

In our story, a clear morning now dawns. All the clouds are dissipated as if by magic. Humankind discovers the Sun and the stars. These luminaries are not fixed in the celestial vault, but move across it periodically. Foucault realizes then that the period of their movement coincides closely with that of the plane of his pendulum. He amuses himself by starting the pendulum in such a way that the Sun lies in the oscillatory plane. As the Sun moves in the sky, the plane of oscillation turns as if to remain oriented toward it. Could it be that the Sun attracts the pendulum and locks the oscillatory plane onto its direction?

Foucault improves the suspension of his pendulum, allowing him to prolong the length of the experiment. He then finds that the Sun drifts slowly relative to the plane of the pendulum. After a month, it is fully 15 degrees off. It appears that the oscillatory plane turns a little more rapidly than the Sun. Meanwhile, astronomers have been busy cataloging the stars in the night sky. Foucault, after several tries, decides to orient his pendulum relative to some brilliant star—Sirius, for example—rather than the Sun. The result is clearly better. Sirius remains in the oscillatory plane for several months. Vega or Arcturus would work just as well. Is it the ensemble of all the brilliant stars that fixes the orientation of the oscillatory plane? Over the years that follow, Foucault realizes that even the brilliant stars are not faithful to him. Slowly but inexorably, each of them drifts relative to the plane of the pendulum. This does not really astonish him, for the astronomers have already told him that the stars are not fixed in the heavens. They orbit about the center of the galaxy. Should he then choose the center of the galaxy, or go even farther? Should he take the Magellanic Clouds as his reference? The Andromeda Nebula? A waste of effort. They all end up (after a long time, it is true) deviating from the oscillatory plane. Foucault notes, however, that the farther away a galaxy is, the longer is its period of fidelity. He thus discovers that, by choosing a number of galaxies many billions of light-years away as his reference, he can finally obtain a stable alignment. (For the expert, I must confess that I have neglected a fully understood general relativistic effect.)

In other words, the Foucault pendulum, aware of the hierarchy of cosmic masses, "ignores" the presence of our little planet, despite its proximity, to align its motions relative to the brotherhood of galaxies that contain most of the matter in the universe. But these galaxies represent almost the entirety of the observable universe. Everything happens *as if the oscillatory plane were constrained to remain fixed relative to the universe as a whole.* (In technical terms, we express the problem in the following way. Among all the systems in relative rotation, there is one in which projectiles in free fall travel in straight lines. This is called an "inertial" system. We may define a second system by requiring that it remain fixed, that is, that it not partake of rotational motion, relative to an ensemble of distant galaxies. Then, by experiment, we find that these two systems coincide. Why?)

The discovery of the fossil glow has provided a magnificent corroboration of this statement. This radiation was emitted 15 billion years ago, when the universe was a thousand times as hot as it is today, and no stars or galaxies yet existed. We can measure the rotation, relative to us, of the layer of matter 15 billion light-years away that emitted the fossil radiation we now receive on Earth. *This source is found to be stationary relative to the plane of oscillation.* (The fossil glow is not exactly the same in all directions. But the deviation varies with the angle of our observation. This variation shows that the deviation arises from the motion of the Earth itself, not from a rotation of the source relative to the oscillatory plane.)

How can we explain the behavior of the pendulum? At the end of the last century, the Austrian physicist Ernst Mach (of supersonic speed fame) tried to see in it the presence of a mysterious influence that emanated from the mass of the universe in its totality. No one has yet gone very far in exploring this hypothesis, which is called "Mach's

Principle." Some physicists have even criticized the label "principle." I agree; I prefer to view it as a seductive intuitive insight, but one difficult to follow up and use.

Where do the laws of physics come from? Their very existence is profoundly mysterious. What hidden power commands electric charges to attract or to repel? On what stone tablets are engraved the modes of interaction of elementary particles? Mach has perhaps lifted a corner of the veil. The "force" that orients the oscillatory plane is born from the universe's "global" action on the "local" motion of the pendulum. The same scheme is perhaps applicable to all the forces of physics. Mach's insight becomes a sort of research plan, a new direction to pursue. (The recent evolution of research in the physics of elementary particles seems to point in this direction. We have begun to relate the various forces of nature within a single framework. We can then invoke the behavior of the universe as a whole, especially its expansion, to explain how these forces came to differ.)

Mach has left us with this realization: The entire universe is mysteriously present at every place and every instant. This leads us very far indeed from traditional concepts of matter, time, and space.

The Law Is the Same Everywhere

The Foucault pendulum has suggested to us a kind of omnipresence of matter, or at least of its influence. Even though located at a mean distance of several billion light-years, it constrains the plane of the pendulum to remain fixed in space, despite the rotation of the Earth. We shall now study two observations, of a completely different type, which are possibly not unrelated to the phenomena discussed above.

When you go to a country house in wintertime, the first thing you do is light the fire. You know from experience that it will take a while for the house to achieve a uniformly comfortable temperature. Collision by collision, each of the air molecules must receive its share of the heat given off by the fireplace. "Causes," whatever their nature, always take some time to produce their "effects." In the example given here, the transmission is particularly slow. In another case, it could be much faster. My voice, when I call to someone, propagates at about 1000 kilometers per hour. But modern physics imposes a limit: No effect can propagate faster than light. Several years ago, radio astronomers sent some messages into space, in the hope that they would be received and that answers would come back. But they knew they would need a lot of patience. No signal could reach even the nearest star in less than four years, or the Andromeda Nebula in less than 2 million years.

It is physically impossible to communicate *today* with the inhabitants of Andromeda.

To describe this situation, physicists use the expression, "causally related." We are not causally related to Andromeda today (that is, Andromeda cannot feel today the effects of a cause that takes place today on Earth), but we are causally related to Andromeda in 2 million years. (This use of the present tense—"we are"—is not an error. *We are today* causally related to Andromeda as it *was* 2 million years ago and as it *will be* 2 million years from now. This point is fundamental.) Equipped with this idea, we shall now return to the observations of the "fossil glow." One very important characteristic of this glow is its isotropy; to very great precision (about one part per thousand), its temperature is the same in all directions. It was, we recall, emitted a million years after the origin of the universe by an ensemble of atoms that are now located about 15 billion light-years from us.

Let us aim our telescope first to the east and then to the west. We can show that the atoms that gave off the fossil glow coming from the east were not (and never have been) causally related to the atoms that emitted the fossil glow coming from the west. This gives rise to the question that preoccupies astrophysicists today: How can regions of the sky that, since the origin of the universe, have never been causally related to each other have exactly the same temperature? How were the orders sent? (*N34*)

Here is another observation that is perhaps even more mysterious. Laboratory experiments tell us that atoms emit radiation with very precise wavelengths. Atoms of hydrogen, for example, can emit radiation with a wavelength of 21 centimeters (among other wavelengths). Analysis of the light coming from stars permits us to identify the atoms at their surface and thus to learn their chemical composition. Physics tells us why an atom emits photons of a given color rather than some other color. Without going into details, we can say that the process reflects the action of the electromagnetic force upon the particles that make up the atom. It is, to be precise, the strength of this force, for a given atom, that permits the emission of one wavelength and forbids another. We have observed radiation emitted from quasars separated by billions of light-years. By comparing their radiation, we have shown that in all these sources the electromagnetic force that governed the emission of the light was exactly the same. Any difference in strength, however small, would have had observable effects on the wavelengths received at Earth. But we have reasons to believe that, at the moment when the quasars emitted these photons, *they were not causally related to one another.*

In summary, these are objects that all obey the same laws of physics without their respective matter ever having been in communication. As with the temperature of the fossil glow, we must ask how "orders" could have been sent outside the bounds of physical causality.

Do these questions make sense? Is there any reason to pursue them? Some of my colleagues think not. For them, this is "metaphysics." They accept the existence of the laws and their omnipresence as observational data in no need of further explanations. I cannot follow them. I have the impression that physics has arrived at a stage in its evolution where these questions legitimately enter its domain.

Atoms That Stay in Contact

This is a particularly difficult subject. It involves some unusual concepts. Readers who lose their footing will find a short summary a little farther on.

Let us study, in a physics laboratory, an unstable particle with no electric charge, which we shall call 0 (figures 54 and 55). It soon disintegrates into two particles with opposite charges, which we shall call + and −. These particles fly off at great speed in opposite directions. (To simplify the discussion, I have described a fictitious event. Reactions only slightly more complex, containing all the elements we find here, do in fact take place.) At the start, nothing specifies their directions. They may be north–south, east–west, up–down; all these directions are a priori equally probable. Several meters away, let us say in a direction due east of the point of disintegration, we place a detector. The + particle happens to be detected by it. We then deduce that, at the moment of disintegration, this + particle was emitted in the eastward direction. In consequence, we are entitled to conclude that the − particle was emitted toward the west. We can test this experimentally, and we find that everything has happened as predicted.

The problem is that we have used language and arguments that are not well suited to the realm of atoms. We have assumed that each emerging particle had, upon separation, a precisely defined direction (figure 54). Quantum mechanics states, on the contrary, that between the moment of disintegration and the moment of detection, no direction whatsoever was assigned to these two particles. It is the act of detection itself that fixes this property (figure 55).

We might try to ignore this assertion. We could formulate a more "reasonable" theory, based on the idea that the particles *do* have well-defined directions from the moment of disintegration. Laboratory ex-

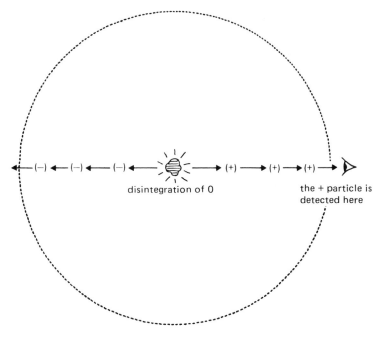

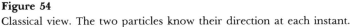

disintegration of 0   the + particle is
                                                       detected here

Figure 54
Classical view. The two particles know their direction at each instant.

periments designed to distinguish between the predictions of quantum mechanics and the "more reasonable" theories have, however, all found in favor of quantum mechanics. In other words, if we reject the idea that the choice of direction is made only at the time of detection, then we predict results that are contrary to actual observation. But if we accept quantum mechanics, we predict results that accord well with experiments.

We can dramatize this situation by imagining that we place one of our detectors in the Andromeda Nebula. Two million light-years separate the disintegration on Earth and the detection. Yet we have every reason to believe that the second particle learns instantly the properties it must now have. Of course, this experiment has not been carried out, but quantum mechanics leaves no ambiguity on the subject. Here, then, is a crucial question: How does the undetected − particle, which does not know its direction before the + particle is detected in the east, know that it "ought" to be heading westward? This enigma bears, in the specialized literature, the name "EPR (Einstein–Podolsky–Rosen) paradox." It was formulated to try to show a flaw in quantum mechanics. The enigma still remains.

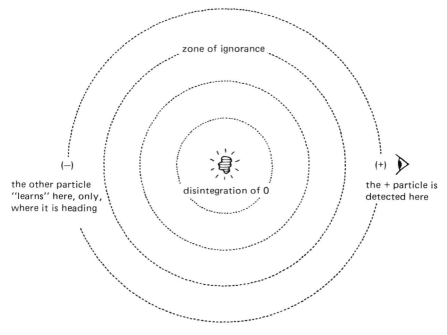

Figure 55

Quantum view. Between the moment of disintegration and the moment of detection, no direction can be assigned to the particles.

I resume our narrative with a comparison aimed at the reader who is less familiar with atomic physics. We give the following assignment to two messengers traveling in opposite directions: They must answer a question with a "yes" or a "no." If the first answers "yes," the second must answer "no," and vice versa. Things work out just as expected. It would be reasonable to suppose that they were both given the word before departing, and that at each instant along their way each one knew what the other would answer. We can show, however, that this is not the case: Neither messenger chooses an answer to give until the last moment. How, then, can we explain the fact that the second messenger knows the right answer?

According to a good number of physicists (but not all of them), the situation can be clarified in the following manner. The two particles (or the two messengers) form a system that must be considered as a whole, whatever the dimensions of the system may be. The solutions of many of the paradoxes of physics (or of science in general) have required the rejection of prejudices unanimously held and regarded as "obvious" by all researchers. Here the prejudice is that the properties of particles are "localized" within the system. The paradox arises from the supposition that information is localized in the particle and must "propagate" in order to get from one particle to the other. Quantum mechanics implies, on the contrary, that the two particles remain in permanent contact however far apart they might be, even if they are no longer causally related. From this fact, we see that no information is obliged to travel from one particle to the other.

We might inquire whether there is any relationship between the problem of the temperature of the fossil light and the problem posed by the EPR paradox. The second case involves only two particles; the first, all the particles in the observable universe (at the time when the cosmic clock read 1 million years). We are tempted to explain the homogeneity of the temperature at that moment by appealing to such a permanent contact between particles, outside the framework of normal causality. The uniformity of the laws of physics would also be related to this same property of matter. In a sense, all of the universe is always and everywhere "present" to every part of itself.

In these three cases—Foucault's pendulum, the observation of distant objects, and the detection of particles—we have been led to ascribe to matter certain influences that lie quite beyond those to which we are accustomed. It seems that there might be two levels of contact between things. First would be the level of traditional causality. Second would be a level that involved neither the force of one body on another, nor any exchange of energy. Rather, it would seem to involve an immanent and omnipresent influence that is difficult to describe with precision (*N30*). I would very much like to know what connections might exist between this hypothetical influence and cosmic evolution.

Appendixes

It is not easy to speak of light in a way that is both simple and correct. I shall present here what one needs to know about light in order to understand the preceding pages, without worrying about rigor.

An insect stirs on the water of a tranquil pond. All around it, circular waves spread outward and propagate to the banks. The distance between two crests is the "wavelength." The number of crests that arrive at the edge each second is the "frequency" of the wave. The higher the frequency, the shorter the wavelength. The sounds that emerge from a loudspeaker are also waves. It is the air that vibrates. These are not concentric circles, but concentric spheres that propagate outward in all directions. Their speed is about 300 meters per second (700 miles per hour). The wavelengths range from several meters for the lowest-pitched sounds to several centimeters for the highest pitches; the frequencies range from several tens to several thousands of cycles per second.

Waves of light emerge from a candle that burns in the night. Like sound waves, these are concentric spheres. Yellow light from the candle has a wavelength of about half a micron (a micron is one ten-thousandth of a millimeter, about the thickness of a soap bubble). The crests arrive at our eyes at the speed of light with a frequency of about 600 trillion (6×10^{14}) cycles per second. One could equally well describe the phenomenon by saying that the candle emits particles of light (photons) in all directions. These photons travel at the speed of light over the distance between the candle and our eyes.

How can we reconcile these two representations?

We say that the energy of the photons is proportional to the frequency of the wave. Yellow light waves are associated with photons that each have an energy of about 2 electron volts. Violet waves correspond to photons of 4 electron volts. By varying the wavelengths (or the energy of the photons), we can cover the entire range of

Table 1

The electromagnetic spectrum and its subdivisions. The greek letter μ (mu) denotes a micron (one thousandth of a millimeter). On the energy scale, we have also indicated the general strength range of electromagnetic and nuclear bonding.

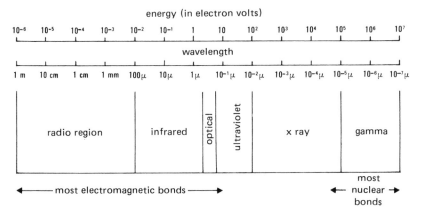

electromagnetic waves. From a kilometer to a centimeter, we are in the radio range. At shorter wavelengths, we pass to millimeter waves (used in radar), then to the infrared, which extends to about 1 micron. Visible light occupies a narrow band between 0.7 and 0.4 micron. Finally, we encounter in turn ultraviolet, x, and gamma rays. Table 1 shows the frequencies and energies of photons corresponding to each of these types of radiation.

How do we define the temperature of luminous radiation? Let us heat a furnace to 1000 degrees centigrade. The walls turn red. The cavity fills with red radiation, which we say is "at 1000 degrees." The yellow light of the Sun corresponds to a temperature of 6000 degrees, and the blue color of Vega arises from its temperature of 12,000 degrees. The center of the Sun, at 16 million degrees, is a site of intense gamma radiation. The 3-degree radiation (the fossil glow) is what would fill a cavity cooled to this temperature. It is invisible to the naked eye but can be detected by radio telescopes. It has a mean wavelength of 3 millimeters and a mean energy of 3 ten-thousandths of an electron volt.

Thanks to quantum mechanics, the mysterious behavior of light is now clear. Unfortunately, it is not possible to explain it further within the scope of this book (it would first be necessary to cover a large amount of mathematics). It turns out that the notion of waves and particles, which are well adapted to reality on our normal scale, are much less appropriate at the atomic level. We would nonetheless be almost correct in saying that light sometimes behaves like a train of waves, and sometimes like a jet of particles.

Day and night, we receive from the Sun about 10 billion neutrinos per square centimeter. Like the electron and the proton, the neutrino is an elementary particle. It has no electric charge. Its mass, if it is not (perhaps) zero like that of a photon, is at least 10 thousand times less than that of the electron (less than 50 electron volts).

In the terminology of modern physics, the electron and the neutrino belong to the same family, that of leptons (or lightweight particles). In our context, the most important property of neutrinos is the weakness of their interaction with the rest of the universe. A thin sheet of paper suffices to intercept photons of visible light. A lead screen capable of stopping neutrinos would have to be several light-years thick. The Sun sends us neutrinos even at night, since they penetrate the volume of our planet without any difficulty. Hundreds of tons of detectors are needed to stop just a few of them. For this reason, it was not until 1954 that neutrinos were actually detected.

But physicists already knew about neutrinos. Wolfgang Pauli had postulated their existence in 1931 to explain certain peculiarities in the disintegration of neutrons. It seemed that energy was not being conserved, nor were certain other quantities, during this process. The new particle was assigned the role of saving the situation. After this rather timid entry upon the scene, neutrinos have assumed a more and more important place in physics and cosmology. They play a fundamental role at the level of the structure of matter. They dominate the behavior of stars in their old age. They may also govern the expansion of the universe.

Theory predicted, in addition to the fossil glow, the existence of a universal population of neutrinos. Like photons, they were believed to be without mass. Recent experiments have raised the possibility that neutrinos might have a tiny mass, which could increase the mean density of the universe up to the critical level necessary to close it. This does not seem very likely, however, at the present time (1983).

Our problem here is that we have run up against some very difficult physics, that of high energies. The answers will come from experiments carried out in laboratories near giant accelerators, such as the European Center for Nuclear Research (CERN) in Switzerland. We have already learned many facts about the physics of neutrinos. We now know of three different species, each of which played a specific role in the behavior of the universe at the time of its origin. How many other species exist? What is their influence on the cosmos? Today cosmology is studied in the irradiation chambers of accelerators as well as in astronomical observatories.

Our first genealogical tree

I would like to set out, in a more systematic way, the different structural elements or "particles" that play a role in our universe. These elements cluster naturally into families. Combinations of members of a family form the elements of a new and larger family; and each such family corresponds to one level of cosmic complexity.

Nature, with its habitual prodigality, created many more particles than it really uses. At least, that is how it seems to us. Some of their roles still escape us. To keep my story short, I shall limit myself to the ones that appear to count.

The lifetime of a particle is an important factor in determining the roles it can play. The photon (or particle of light) is apparently eternal. It is present at all levels of reality, as much at the moment of formation of nuclei as in warming your soul when you watch a sunset.

We presently think that quarks have a limited, albeit very long, lifetime (10^{32} years). This assures the universe sufficient stability to elaborate the complexity of the cosmos. Conversely, other particles, with lives measured in millionths or billionths of a second, seem to represent only intermediate steps in transitions to more stable states.

We shall review, level by level, the plans of the structure of matter. At each step we shall compile an inventory of constituents and discuss their demography on the cosmic scale.

This enumeration will be carefully conducted, since it deserves our close attention. We are, in a sense, constructing our own genealogical tree.

The Level of Quarks and Leptons

At the most elementary level of reality, five principal actors dominate the scene; two quarks with electric charges of $+2/3$ (U quarks) and

−1/3 (D quark), the electron and the neutrino (collectively called leptons), and the photon. All visible matter, from our planet to the most distant quasar, is made of a combination of U and D quarks and electrons. Exchanges of energy are carried out for the most part by photons and, to a lesser extent, by neutrinos.

Contemporary physics would describe the situation in the following terms. "Matter" is composed of quarks, electrons, and neutrinos, while "forces" are carried by a collection of "exchange" particles known as photons (for electromagnetism), intermediate bosons (for the weak force), gluons (for the strong nuclear force), and gravitons (for gravitation). Of these exchange particles, photons were the only ones to have been detected until very recently. The discovery of intermediate bosons was reported at CERN in 1983.

Photons are the most abundant particles. There are about 400 per cubic centimeter. The large majority belong to the fossil glow. The neutrino population is not well known but is expected to be similar to the photon population. Here again the vast majority belong to a radiation of fossil neutrinos born in the first few seconds of the universe.

The populations of quarks and electrons are practically equal, a little less than one per cubic meter—a billion times smaller than the abundance of photons or neutrinos.

A more precise calculation gives 2.14 U quarks and 1.28 D quarks per electron.

The Level of Nucleons

Let us move on to a higher level. Here the quarks combine in threes to produce the vast family called "hadrons" (or strongly interacting particles). Their lifetimes are generally measured in trillionths of a second. They disintegrate into protons or neutrons. An isolated neutron will disintegrate into a proton in about 15 minutes, but once incorporated into a nucleus, it becomes as stable as its relative the proton.

The photon is made of two U quarks and one D quark (a total charge of +1), while the neutron is made of two D quarks and one U quark (a total charge of zero). An inventory of nature shows that there are as many protons as electrons, and five times as many protons as neutrons. The overwhelming majority of the neutrons that exist today are incorporated into helium nuclei.

The Level of Atomic Nuclei

Nucleons cluster into nuclei. Then electrons enter into orbits about the nuclei to form atoms (table 2). The number of protons (or atomic

Table 2

Several simple atoms.

•	●	○
Electron	Proton (hydrogen nucleus)	Neutron

atom of light hydrogen

atom of heavy hydrogen (deuterium)

atom of helium–3

atom of helium–4

atom of carbon–12

atom of oxygen–16

number) is equal to the number of electrons, and this number determines the chemical nature of the atom. An atomic number of 1 denotes hydrogen, 2 helium, 3 lithium, and so on. Atoms with atomic numbers higher than 83 are unstable.

The atomic elements frequently have several different forms or "isotopes." When we discussed the age of the universe, for example, we spoke of carbon-14. The number 14 indicates that the nucleus of the atom consists of 6 protons and 8 neutrons, giving 14 nuclear constituents or "nucleons." Carbon is defined by the presence of 6 protons, independent of the number of neutrons. Two other isotopes of carbon exist in nature: carbon-12 (6 protons and 6 neutrons) and carbon-13 (6 protons and 7 neutrons). Both differ from carbon-14 in that they are stable. The carbon in our bodies, like that in the solar system in general, is about 99 percent carbon-12 and 1 percent carbon-13.

Each chemical element from hydrogen to uranium, when subjected to appropriate excitation (in a fluorescent bulb, for example), emits light made up of a superposition of a number of "rays" of precisely defined colors, called a "spectrum." By means of a prism or spectroscope, the physicist can decompose this light and identify the element by its spectrum, just as individuals can be identified by their fingerprints. By analyzing the light of a star with a spectroscope, the astronomer can not only determine which chemical elements are present but also estimate the relative abundances of the elements at the star's surface. Thanks to this method, which has now become highly sophisticated, we have a good idea of the chemical compositions of stars and galaxies.

The Abundances of the Elements in the Universe

Stellar and interstellar gestation of the chemical elements took place over a span of 15 billion years. In table 3, the results of this gestation are displayed in the form of the relative abundances of the elements. The table gives mean values, valid both for the universe as a whole and for our Sun, for the stars that surround us, and for external galaxies. There are some differences among these objects, but they are relatively minor. However, the table does not apply to planets or other small bodies.

We have arbitrarily fixed the abundance of hydrogen at a trillion (10^{12}) atoms (data compiled by J.-P. Meyer and A. G. W. Cameron). Elements 43, 62, 84–89, 91, 93, and higher have only unstable isotopes with relatively short lifetimes. Their abundances in nature are too small to be measurable.

Table 3
Abundances of the chemical elements in the universe.

Name	Symbol	Number of protons or electrons	Relative abundance
Hydrogen	H	1	1.0×10^{12}
Helium	He	2	8.5×10^{10}
Lithium	Li	3	1000
Beryllium	Be	4	15
Boron	B	5	200
Carbon	C	6	4.8×10^{8}
Nitrogen	N	7	8.5×10^{7}
Oxygen	O	8	8.0×10^{8}
Fluorine	F	9	3.4×10^{4}
Neon	Ne	10	1×10^{8}
Sodium	Na	11	2.1×10^{6}
Magnesium	Mg	12	3.9×10^{7}
Aluminum	Al	13	3.1×10^{6}
Silicon	Si	14	3.7×10^{7}
Phosphorous	P	15	3.5×10^{5}
Sulfur	S	16	1.7×10^{7}
Chlorine	Cl	17	1.7×10^{5}
Argon	Ar	18	3.3×10^{6}
Potassium	K	19	1.3×10^{5}
Calcium	Ca	20	2.3×10^{6}
Scandium	Sc	21	1.3×10^{3}
Titanium	Ti	22	1×10^{5}
Vanadium	V	23	1×10^{4}
Chromium	Cr	24	4.8×10^{5}
Manganese	Mn	25	2.9×10^{5}
Iron	Fe	26	3.3×10^{7}
Cobalt	Co	27	7.8×10^{4}
Nickel	Ni	28	1.8×10^{6}
Copper	Cu	29	1.9×10^{4}
Zinc	Zn	30	5×10^{4}
Gallium	Ga	31	1800
Germanium	Ge	32	4300
Arsenic	As	33	240
Selenium	Se	34	2500
Bromine	Br	35	520
Krypton	Kr	36	1700
Rubidium	Rb	37	220
Strontium	Sr	38	1000

Table 3 (continued)

Yttrium	Y	39	185
Zirconium	Zr	40	1000
Niobium	Nb	41	52
Molybdenum	Mo	42	150
Technetium	Tc	43	—
Ruthenium	Ru	44	70
Rhodium	Rh	45	15
Palladium	Pd	46	48
Silver	Ag	47	17
Cadmium	Cd	48	56
Indium	In	49	7.4
Tin	Sn	50	130
Antimony	Sb	51	11
Tellurium	Te	52	240
Iodine	I	53	41
Xenon	Xe	54	200
Cesium	Cs	55	15
Barium	Ba	56	180
Lanthanum	La	57	15
Cerium	Ce	58	44
Praesodymium	Pr	59	5.5
Neodymium	Nd	60	30
Promethium	Pm	61	—
Samarium	Sm	62	7.4
Europium	Eu	63	3.0
Gadolinium	Gd	64	11
Terbium	Tb	65	1.8
Dysprosium	Dy	66	13
Holmium	Ho	67	3.0
Erbium	Er	68	7.4
Thulium	Tm	69	1.3
Ytterbium	Yb	70	8.1
Lutecium	Lu	71	1.5
Hafnium	Hf	72	7.8
Tantalum	Ta	73	0.7
Tungsten	W	74	5.9
Rhenium	Re	75	1.8
Osmium	Os	76	28

Table 3 (continued)

Iridium	Ir	77	27
Platinum	Pt	78	52
Gold	Au	79	7.4
Mercury	Hg	80	15
Thallium	Tl	81	7.0
Lead	Pb	82	111
Bismuth	Bi	83	5.2
Polonium	Po	84	—
Astatine	At	85	—
Radon	Rn	86	—
Francium	Fr	87	—
Radium	Ra	88	—
Actinium	Ac	89	—
Thorium	Th	90	1.8
Protoactinium	Pa	91	—
Uranium	U	92	1.1
Neptunium	Np	93	—
Plutonium	Pu	94	—
Americium	Am	95	—
Curium	Cm	96	—
Berkelium	Bk	97	—
Californium	Cf	98	—
Einsteinium	Es	99	—
Fermium	Fm	100	—
Mendelevium	Mv	101	—
Nobelium	No	102	—
Lawrencium	Lw	103	—

Table 4
Relative abundances of the elements (percentage numbers of atoms)

Universe		Earth's crust		Seawater		Human body	
H	90.	O	47.	H	66.	H	63.
He	9.	Si	28.	O	33.	O	25.5
O	0.10	Al	8.	Cl	0.33	C	9.5
C	0.06	Fe	4.5	Na	0.28	N	1.4
Ne	0.012	Ca	3.5	Mg	0.033	Ca	0.31
N	0.01	Na	2.5	S	0.017	P	0.22
Mg	0.005	Mg	2.2	Ca	0.006	Cl	0.03
Si	0.005	Ti	0.46	K	0.006	K	0.06
Fe	0.004	H	0.22	C	0.0014	S	0.05
S	0.002	C	0.19	Br	0.0005	Mg	0.01

Hydrogen predominates, comprising 90 percent of the atoms in the universe. Less than one atom per thousand is neither hydrogen nor helium. Quantitatively, the stars have not synthesized very much. But qualitatively, their contribution has been sufficient to launch the universe upon the paths of complexity.

Cosmic evolution occurs in an ever more specialized medium. It is interesting to see how the composition of these settings differs from the mean cosmic values. Table 4 presents the most abundant elements (by percentage) in the universe, in the terrestrial crust, in seawater, and in the human body. The order is not constant. Hydrogen, carbon, and nitrogen, for example, are very poorly represented in the Earth's crust. The fact that H and O are at the top of the list in the human body reminds us of the importance of water (H_2O) for the building of large structures. The fact that chlorine, sodium, magnesium, and potassium appear in the last two lists tells us that life originally developed in the oceans. The differences between the two lists are especially great for carbon and nitrogen, which are privileged elements, along with hydrogen and oxygen, in the structures of the largest molecules.

Some Fruits of Chemical Evolution

The chemical activity that occurs in clouds of interstellar matter results in the buildup of a number of "interstellar molecules." Table 5 lists those identified prior to 1981. They are grouped according to the number of atoms they contain.

Table 5
Interstellar molecules discovered and identified up to early 1981.

Number of atoms	Chemical formulas
2	H_2, CH, OH, C_2, CN, CO (carbon monoxide), NO, CS, SiO, SO, NS, SiS
3	H_2O (water), C_2H, HCN, HNC, HCO, NH_2^+, H_2S, HNO, OCS, SO_2
4	NH_3 (ammonia), C_2H_2 (acetylene), HNCO, C_3N, H_2CO (formaldehyde), HNCS
5	CH_4 (methane), CH_2NH, CH_2CO, NH_2CN, HCOOH (formic acid), C_4H, HC_3N
6	CH_3OH (methyl alcohol), CH_3CN, NH_2CHO, CH_3SH (methyl mercaptan)
7	CH_3NH_2, CH_3C_2H, CH_3CHO, CH_2CHCN, HC_5N
8	$HCOOCH_3$
9	CH_3CH_2OH (ethyl alcohol), CH_3CH_2CN, HC_7N, $(CH_3)_2O$
11	HC_9N
13	$HC_{11}N$

The Principal Actors in Biological Evolution

Among the great diversity of molecules that were built up in the primitive ocean, two groups attract our attention: the amino acids, which string together to form proteins, and the nucleotide bases, which unite to form DNA. Table 6 lists the twenty amino acids used in the transmission of the genetic code and shows how their constituent atoms are arranged.

Let us briefly mention some details regarding the typography of the genetic code. The letters A, C, G, and T designate the four nucleotide bases adenine, cytosine, guanine, and thymine (table 7). Genes are long molecular chains in which the nucleotide bases form the links. The entire chain, called DNA (deoxyribonucleic acid), is arranged in the form of a double helix. Models are displayed in most science museums.

There is a convention of nature that, to each triplet of bases, there corresponds a particular amino acid. For example, the triplet AGT corresponds to valine, and the triplet TGC to aspartine. In the ribosomes of the cell, proteins serving vital functions are assembled from amino acids. The choice of these amino acids is fixed by the sequence of bases in the DNA molecule. Each animal, each individual, has a dis-

Table 6
The amino acids.

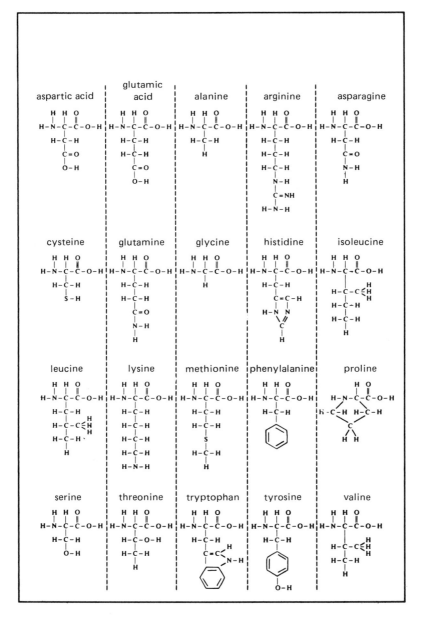

Table 7
The nucleotide bases.

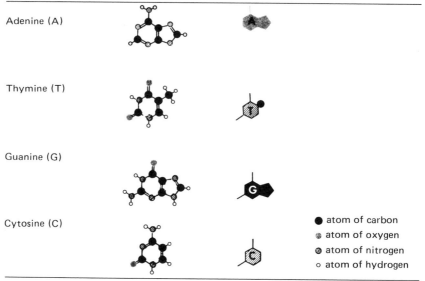

Adenine (A)		
Thymine (T)		
Guanine (G)		
Cytosine (C)		● atom of carbon
		⬤ atom of oxygen
		⊘ atom of nitrogen
		o atom of hydrogen

tinctive sequence, thanks to which food does not cause mutations and offspring resemble their parents.

The Architecture of a Protein

The sequence and spatial arrangement of the amino acids in a protein determine the precise nature of each of the countless functions that vegetable and animal life requires. Figure 56 shows a molecule called "cytochrome-C," which is extracted from the heart of a horse. Its role is to capture, and to introduce into the blood circulation, molecules of oxygen supplied by pulmonary respiration. Composed of the twenty amino acids described above, it is illustrative of the spatial structure of proteins.

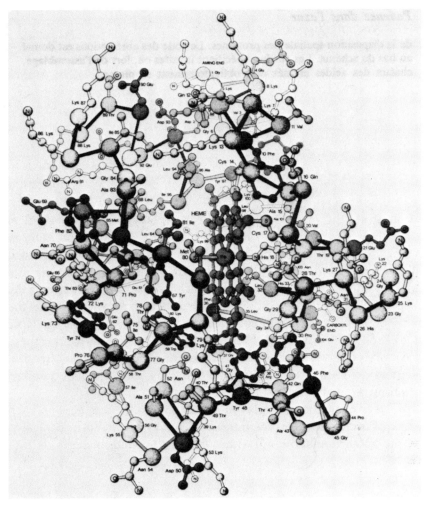

Figure 56
The molecule cytochrome-C. After Richard E. Dickerson (*Scientific American*, April 1972). The numbers specify where in the sequence each of the amino acids must be placed.

Key:

Asp	aspartic acid	Leu	leucine
Glu	glutamic acid	Lys	lysine
Ala	alanine	Met	methionine
Arg	arginine	Phe	phenylalanine
Asn	asparagine	Pro	proline
Cys	cysteine	Ser	serine
Gln	glutamine	Thr	threonine
Gly	glycine	Trp	tryptophan
His	histidine	Tyr	tyrosine
Ile	isoleucine	Val	valine

To illustrate the major steps in nuclear evolution, it is convenient to place the atomic nuclei in a large array (figure 57) of proton number (vertical) versus neutron number (horizontal). Here we have displayed all the stable isotopes up to silicon (plus two unstable species: the free neutron and carbon-14). The stable nuclei are distributed close to the diagonal line, along which the number of neutrons exactly equals the number of protons. For higher atomic numbers, the zone of stable nuclei moves slowly from that diagonal into the region where there are more neutrons than protons. The isotopes of a particular chemical element are located on a horizontal line (they have the same number of protons but different numbers of neutrons).

Nuclear evolution is displayed in this format (figures 58–65). Each phase of evolution effects the transformation of certain nuclei (indicated by the hatched squares) into other nuclei (the white boxes).

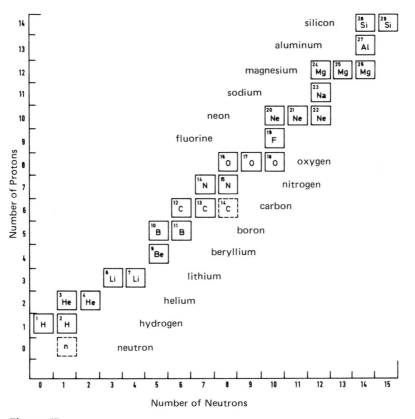

Figure 57
The chart of the nuclides.

I

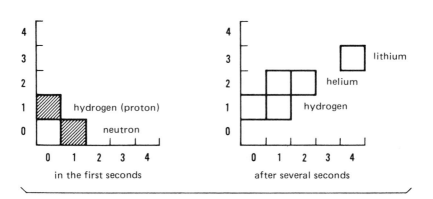

In the Initial Explosion

Figure 58
Nuclear evolution in the initial explosion. Issuing forth from the great brilliance in which the universe was born, protons and neutrons interacted. Several minutes after its birth, the universe was made of hydrogen, helium, and lithium-7. These are the oldest atoms in the universe.

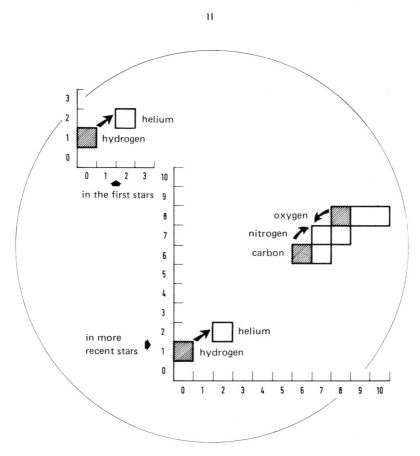

Figure 59

Nuclear evolution in main sequence stars. These stars obtain their energy from the fusion of hydrogen into helium. The first stars of the galaxy, lacking heavy atoms, directly transformed hydrogen into helium. More recent stars can carry out this fusion in a more efficient way, which involves the transformation into nitrogen of carbon and oxygen formed by earlier generations of stars. This new fusion scheme uses carbon nuclei as catalysts, in a process called the "Bethe cycle."

III

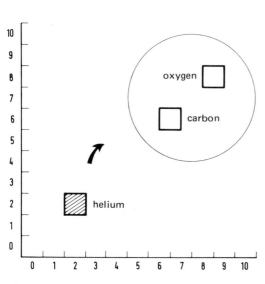

Figure 60
Nuclear evolution in red giants. Helium in the core is transformed into carbon-12 and oxygen-16. In an outer layer surrounding the central core, the fusion of hydrogen into helium goes on just as in the cores of main sequence stars.

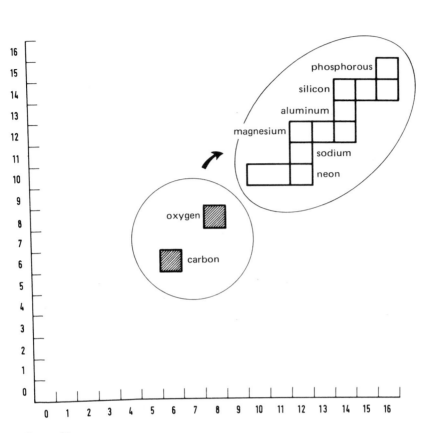

Figure 61
Nuclear evolution in the later stellar stages. The fusion of carbon and oxygen into neon, sodium, magnesium, aluminum, and silicon takes place in the interior of a star during these stages. In outer layers, helium burns. In layers still further out, hydrogen burns.

V

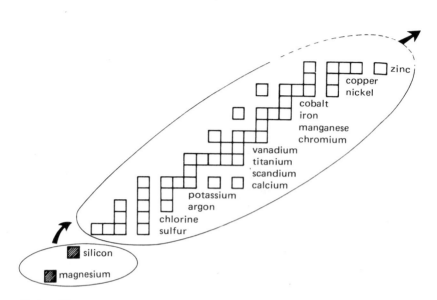

Figure 62

Nuclear evolution in its final stages. Before the star explodes as a supernova, magnesium and silicon fuse to make the metals chromium, manganese, iron, cobalt, nickel, copper, zinc, and so on. Neutrons produced by these reactions combine with the metals to complete the table of the chemical elements.

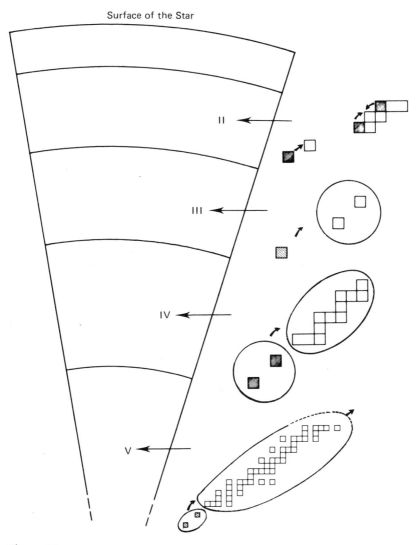

Figure 63
Profile of nuclear transmutations in the interior of a massive, highly evolved star. The reactions that require the highest temperatures (fusion of silicon and magnesium) occur at the center of the star. Farther from the center, the temperature decreases steadily out to the surface, which is much too cold for the play of nuclear reactions.

VI

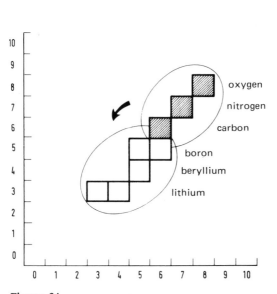

Figure 64

Nuclear evolution in interstellar space. Bombarded by the very fast particles of the cosmic rays, atoms in the interstellar medium are split apart, leaving as a residue nuclei of lithium, beryllium, and boron. To make a single gram of the boric acid we buy at the pharmacy, it is necessary to collect all the boron atoms formed since the birth of the galaxy in a volume of interstellar space as large as our Sun. It is one of the slowest processes in our universe.

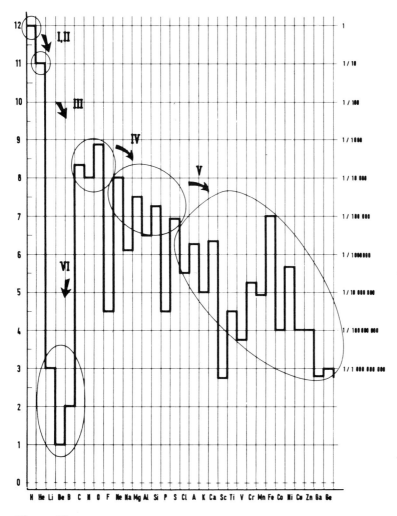

Figure 65

The abundances of the first 33 elements. On the left-hand scale, hydrogen has
an abundance of 10^{12} (1,000,000,000,000) atoms. On the right-hand scale, hydrogen is assigned an abundance of 1. The arrows trace the major phases of nuclear
evolution already described. (Note: The scales are logarithmic. The number 4 on
the left-hand scale means $10^4 = 10,000$; 0 means $10^0 = 1$; and so on.)

This game can be played in two ways. First, we can place all the stars of the sky on a large chart (called the Hertzsprung–Russell diagram; see figures 66 and 67) according to their color (horizontal) and the total intensity of their emitted light (vertical). The stars are not distributed uniformly over the diagram but cluster in certain privileged locations. The most densely populated region—the "main sequence"—lies along a diagonal line. All the stars that burn hydrogen to helium are found in this region, burning at a rate that varies with position along the diagonal. Small red stars, in the lower right, reside there many tens of billions of years; the Sun, at the center, 10 billion years; Sirius, even higher, 2 billion years; and Rigel, in the upper left (the domain of the blue giants), only a few million years. The second region of the diagram, in the upper right, is occupied by red giants and red supergiants, like Betelgeuse, Antares, and Aldebaran, which burn helium to carbon and oxygen in their cores. To their left we find the "horizontal branch," where the stars that are more advanced in their nuclear evolution are located. They fuse carbon, oxygen, and silicon. In the extension of this branch, there is a region occupied by planetary nebulae and then, after an abrupt left-hand turn, the white dwarf region.

In our second round of this game, we place a single star on the chart and track its course over its lifetime (figures 68 and 69). This track, called the H–R trajectory, and the speed of movement along the track, depend upon the mass of the star. Immediately after their birth in the collapse of a large interstellar cloud, very luminous red stars—called T-Tauri stars—move rapidly to the left and into the main sequence. They remain there until they exhaust their central hydrogen supplies. Then they move off to the right, exhaust their helium in the red giant zone, and enter the horizontal branch upon reaching the most advanced phases of nuclear fusion. At this stage, very massive stars attain fatidic temperatures—4 or 5 billion degrees—at which they explode as supernovae. Stars of about one solar mass or less run through the horizontal branch, reach the planetary nebula stage, wind across the main sequence, and coast down the slope of white dwarfs, down the dead-end street of stars deprived of their sources of nuclear energy, dying as black dwarfs. (*N35*)

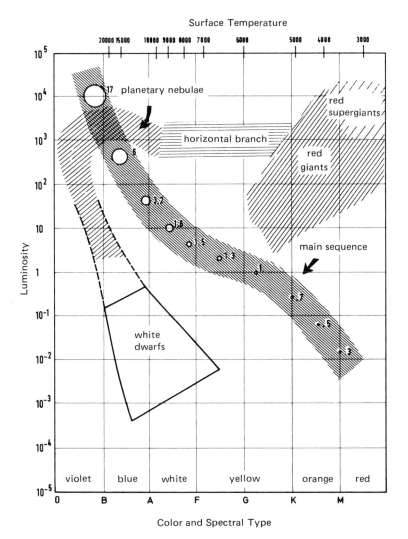

Figure 66

The stellar color-luminosity (Hertzsprung–Russell) diagram. The vertical scale is the absolute luminosity of the star (relative to the luminosity of the Sun), while the horizontal scale gives the color of the star, which is equivalent to a surface temperature scale (top). Astronomers use a more detailed classification scheme of "spectral types," as indicated by the letters O, B, A, F, G, K, and M on the bottom scale. In the hatched region of the main sequence are given the masses of the stars located there, in units of the mass of the Sun.

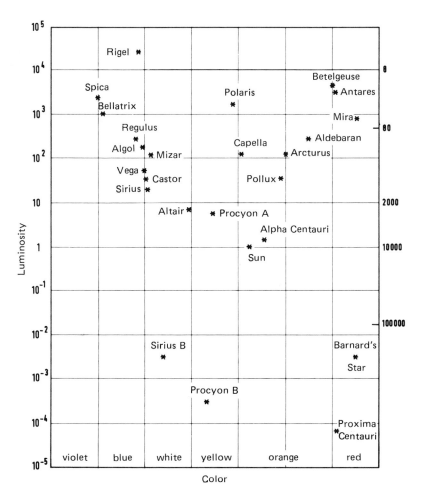

Figure 67

Some familiar stars on the H–R diagram. Each star is placed according to its luminosity and surface color. We can readily see the main sequence, the red giants, and two white dwarfs, Sirius B and Procyon B. The right scale gives, in millions of years, the main sequence residence time, which is the time required to burn all the star's hydrogen into helium. This time represents rather closely the total lifetime of the star, since the later stellar phases are passed through very rapidly. Spica and Bellatrix, for example, will live through all the stages of their lives in a few tens of millions of years, while Barnard's Star or Proxima Centauri will still be shining long after the Sun has become a black dwarf.

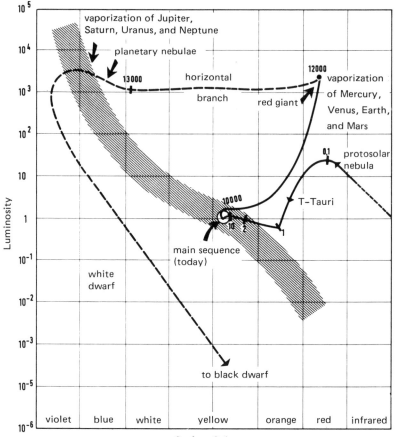

Figure 68

The destiny of the Sun. We have marked on the H–R diagram the trajectory along which the Sun evolves. The numbers give the elapsed time, in millions of years, since the Sun's birth from the collapse of a massive interstellar cloud. After the transformation of the protosolar nebula into a "solar system," and after passing through the T-Tauri phase, it settles down onto the main sequence, where it remains today. Five billion years from now it will chart a course toward the red giant region, then along the horizontal branch. Finally, after passing through the planetary nebula zone, it will slowly die as it follows the path to the white dwarf and black dwarf stages.

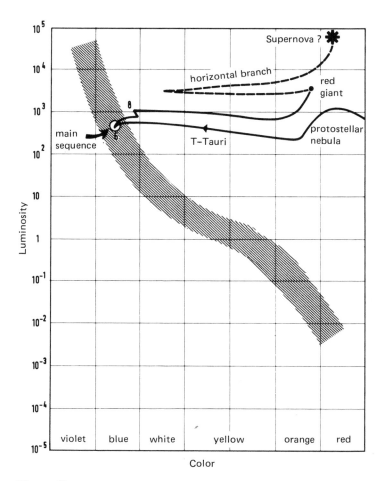

Figure 69

The fate of a massive star. The evolutionary path of a star much more massive than the Sun is traveled at an extremely rapid rate. The star's arrival on the main sequence takes less than a million years, and its residence on the main sequence is ended (as in the case of Rigel) in less than 10 million years. The red giant and horizontal branch stages are even shorter. We suspect, without being certain, that the final supernova explosion occurs somewhere in the red supergiant region.

Although invisible, black holes can be detected. There may be millions of them.

It was Laplace who first came to the idea of black holes. He reasoned as follows. To escape the surface of a planet or star, a projectile must reach a certain minimum speed. This is the escape velocity. Rockets shot off from the Earth must rise at more than 11 kilometers per second; from the Moon, 2 (or more) kilometers per second are required. Let us imagine a celestial body whose escape velocity is greater than 300,000 kilometers per second. Even light could not escape, and the body would thus be invisible. It would be a black hole.

The physical nature of black holes cannot be fully understood without taking into account Albert Einstein's general theory of relativity as a starting point. We might describe what happens by recalling that matter slows the passage of time. More precisely, to a distant observer, time at the surface of a very dense body appears to run slowly. This slowing becomes more pronounced as the density increases. Its effect is to lengthen the wavelength of the light emitted by the body as received by an external observer. Beyond a certain limit, time stops and the wavelength becomes infinite. The wave no longer exists; the light is extinguished.

What is the mass of a black hole? It can have any mass—a gram, a billion tons, or a billion times the mass of the Sun—as long as that mass is concentrated in a small enough volume. A black hole with the mass of Mont Blanc would fit inside the volume of a hydrogen atom. A black hole with the volume of Mont Blanc would have the mass of the Sun.

Even though physical theory permits their existence, nothing proves that black holes do exist in nature. What mechanism could produce such mass concentrations? We know of at least one: the death of a

normal double star

double star, of which one
member is a black hole

Figure 70
Double stars and black holes.

massive star. At the moment of death, stellar matter simultaneously undergoes an explosion in its outer layers (a supernova) and a collapse of its interior. This implosion might give birth to a black hole. Other black holes may have formed amid the extreme densities that prevailed at the beginning of the universe.

How would we detect a black hole? The absence of radiation certainly makes this difficult. We must count on the powerful gravitational field that surrounds such an object. Let us imagine that, tonight, giant hands crush our Sun, reducing its radius to less than a kilometer. The gravity at its surface becomes so intense that no light can escape from it. Tomorrow there will be no sunrise. But the attraction that the Sun exerts on the planets will not be affected. Their movements will not change. The Earth will continue to pursue its annual revolution, as the movement of the constellations in the sky will attest.

There are a great many double stars in the sky. Consider one such pair of very close stars revolving about each other. They "swing" about a point situated between them. If one of the stars happened to be a black hole, we would see only its companion, which would appear to swing all by itself (figure 70). We already know of a few such cases where there may be black holes.

Among the galaxies, quasars are the most powerful emitters of radiation; their output is a thousand times more than a normal galaxy, which is to say, as much as a hundred trillion (10^{14}) Suns. The source of this radiation is concentrated in a minuscule region, no larger than the solar system, at the center of a galaxy. What extravagant mechanism could cause such a vast amount of energy to be emitted from such a small volume? Perhaps a black hole of several million solar masses is to be found there. It may seem paradoxical to invoke the effect of a black hole (from which nothing whatsoever can escape) to explain such brilliant luminosity. But the black hole attracts and devours everything it finds nearby: interstellar clouds, planets, even stars. These bodies, violently accelerated, reach prodigious speeds. The shocks, driven by their infall, heat the bodies up and make them shine brilliantly. Their "swan song" stops abruptly when the maw of the black hole engulfs them irreversibly. The same mechanism may explain the radiation from the cannibal galaxies that occupy the centers of superclusters of galaxies. If this is the case, black holes, far from being rare, may be very numerous. Like stars and galaxies, they would be a common element in the universe.

The study of the properties of black holes continues to produce astonishing results. The English astrophysicist Stephen Hawking has shown that black holes slowly evaporate—a new paradox, since nothing can escape them! It is yet another sleight-of-hand feat of quantum mechanics. The laws are no longer absolute. Tunneling events occur and are responsible for the evaporation. The evaporation of a black hole entails a diminution of its mass, which permits even more rapid evaporation. In the end, it would explode in a fiery flash visible billions of light-years away!

The beginning of the universe, like the heart of the Sun, is hidden from our eyes.

I have spoken of the cosmological horizon. It results from the fact that the most distant galaxies retreat at almost the speed of light. This horizon would exist even if the universe were eternal, as in cosmologies of the "continuous creation" type. It is not related directly to the existence of an initial explosion, but rather to expansion, whatever its cause. In the Big Bang cosmology, a second barrier intervenes. It arises from the fact that, at its origin, the universe was "opaque." It is this opacity at the earliest times that limits our vision of the universe.

To illustrate the nature of this frontier, I shall draw an analogy between the Sun and the universe (figure 71). In this analogy, the center of the Sun will be compared to the origin of the universe, and the radiation from the solar surface will be compared to the fossil glow. We cannot see the center of the Sun because the material of the Sun is opaque, but we can see its surface because the matter situated between us and the surface is transparent. The core of the Sun is its densest and hottest part, at 16 million degrees. The energy produced there is emitted in the form of energetic photons called gamma rays. Solar material is very opaque to these rays. Each photon is emitted, absorbed, reemitted, and reabsorbed a large number of times before reaching the surface. The solar material becomes less and less dense and less and less opaque, the farther we go from the center toward the surface. The trajectory between absorptions becomes longer. A photon takes hundreds of thousands of years to find its way from the center to the surface. From there it travels directly to the Earth in 8 minutes. (The figurative language I use here neglects the fact that photons do not really have individual identities. Furthermore, the degradation in energy entails a proliferation in the number of

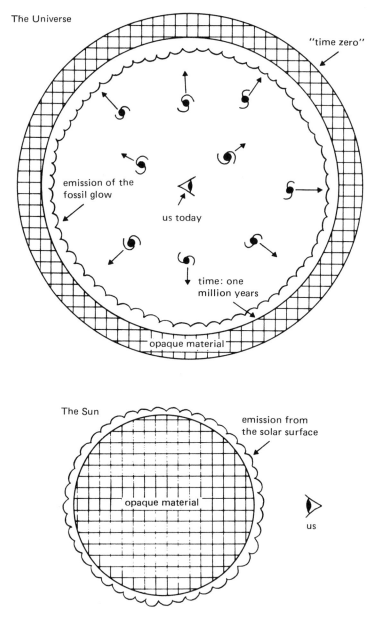

Figure 71

The second horizon (the universe and the Sun). *Above*: A schematic cross section of the universe. The layer of material that emitted the fossil glow 15 billion years ago (inner circle) is now located 15 billion light-years way, receding at 99.9 percent of the speed of light. Material outside that circle is opaque, dense, and hot. Note that this drawing gives the impression that we are at the center of the world. There is, however, no such thing as a "center of the world." Any observers, no matter what galaxy they inhabit, would draw their location in exactly the same way. *Below*: Our Sun as seen from the Earth. We see the solar surface, which emits the radiation we receive. Matter interior to this surface is opaque.

photons.) From this perspective, the solar surface is the place where the photons are mustered out before taking off in our direction.

Let us now imagine a voyage toward the Sun in a fictional capsule constructed of perfectly refractory materials. We see the apparent surface of the Sun grow steadily until it fills half the sky. At that moment, the capsule reaches the surface and penetrates into the Sun. It passes from transparent to opaque surroundings. It moves forward into matter that is ever denser and ever hotter.

In our analogy, the material of the Sun becomes the material of the universe when, at the beginning, it was very dense. Both are opaque. This opaque period of the universe lasted a million years, until the emission of what is now the fossil glow. The space between the Sun and the Earth represents the matter of the universe since the first million years. Both are transparent. The voyage to the Sun symbolizes a trip into the past. Passing through the surface of the Sun is like the moment when the universe passes (going backward) from transparency to opacity. The fossil glow is the analogue to the Sun's visible light. It is made of all the photons that were emitted at the time of the opacity-transparency transition and that have traveled freely since then. These are the oldest photons in the universe. They are as old as the universe, less a million years.

To avoid confusion, I must note an important difference between the two elements of the analogy, the beginning of the universe and the center of the Sun. It is a difference in geometry. The center of the Sun is a point; *the beginning of the universe is everywhere.* In consequence, the photons of the fossil glow reach us from all directions, whereas solar photons reach us from a single well-defined direction. In other words, the matter that emitted the fossil glow has the shape of a spherical shell that surrounds us at a distance of 15 billion light-years, whereas the matter that emits the radiation of the Sun (likewise distributed in a spherical shell) does not surround us. The expansion has degraded these originally red photons to radio waves. It is in this form that Penzias and Wilson detected them.

We have now encountered two horizons in our explorations. The first is due to the recession of the galaxies and is independent of the history of the universe. The second is strictly of a historical nature. It is caused by the opacity of the first moments. The two are at almost exactly the same distance. This is not a coincidence. This is the distance traveled, over the lifetime of the universe, by a galaxy that moves at the speed of light relative to us.

For the knowledgeable reader, here is the formidable situation that confronts the explorer-astrophysicist trying to trace time back to its source.

For one thing, the sphere of causality is shrinking, simply because, the closer we approach time zero (on a linear scale, or time "minus infinity" on a logarithmic scale), the shorter is the propagation time of signals. In parallel, another effect presents itself. According to quantum mechanics, an object cannot be definitively assigned to a location. It has instead what might be called a "sphere of uncertainty of localization." This is the region within which we have the best chance of finding it. We have touched on this in the section devoted to the diamond in the Tower of London. But it happens that, before 10^{-43} second, the sphere of causality was smaller than the sphere of uncertainty of localization. This means that a sample of matter could have found itself in a place with which it was impossible for it to communicate. Understanding here encounters its ultimate horizon; this is the point where physicists throw their hands in the air and say, "I should have been a farmer."

This paradox reveals a profound gap in contemporary physics. No one has yet established a coherent theory that incorporates both Einstein's general theory of relativity and quantum mechanics. No one even knows if such a theory is possible. To conceal this ignorance, the astrophysicist decrees that, at 10^{-43} second, the universe was "born."

Notes

N1. This is a quotation from Baltasar Gracian y Morales by Vladimir Jankelevitch.

N2. See, for example, *Three Ways of Asian Wisdom* by Nancy Wilson Ross (New York: Simon and Schuster, 1966).

N3. The measurement of distances in astronomy: In order to explore and probe our universe, we must first know how to measure distances. In the heavens, I see stars with very different apparent brightnesses. I cannot tell at first sight if one star appears to be brighter than another because it is in fact more luminous, or because it is closer. In the same way, a single light heading toward me on the road at night might belong to a bicycle a short distance off or to a more distant motorcycle. Until I have identified the object, I cannot determine its distance. The art of measuring distances is thus at first the art of determining the nature and intrinsic luminous power of the objects we observe. The light must contain clues that will allow us to identify it. In some cases, the clue is simply color. We know nowadays that one type of blue star shines 100 thousand times as bright as the Sun. In another case, we may use the fact that certain stars are variables; their intensities increase and decrease in a cycle with a well-determined period. The longer the period of variation, the brighter the star. Similarly, the fact that certain stars die in fiery explosions permits us to extend our probing even farther. For a few days, these supernovae attain 100 million times the brightness of the Sun. Even if they are near the limits of the observable universe, they are still visible. Finally, of course, it is necessary to calibrate the standard "candles," that is, to determine their intrinsic luminosities. The techniques are many and complex. To describe them would take us beyond the scope of this book.

N4. Let us note in passing that the numerical value of one billion (10^9) is of profound importance for the unfolding of the universe and its complexity. A slightly larger or smaller value would have resulted in an uninhabitable world. See J. Barrow and F. Tipler, *The Anthropic Cosmological Principle* (New York: Oxford University Press, 1984).

N5. Edgar Morin, *Sciences humaines et Sciences de la Nature*, a France-Culture cassette in the "Connaissance de l'univers" series, October 1978.

N6. A publication of the Centre National de la Recherche Scientifique, 15 quai Anatole-France, 75007 Paris, France.

N7. The old man of the Himalayas: Let us consider a mountain with a base area of about 10 thousand square kilometers and an average height of 10 kilometers (Mount

Everest rises 8 kilometers). The rock has a density of about 3 grams per cubic centimeter. This represents about 10^{44} atoms in the mountain. We know that each atom in the interior of the rock is bound by nearly half an electron volt. The lightest polishing expends some thousands of ergs. (The silk cloth is not the ideal instrument for polishing rock!) We may assume that each rub breaks a hundred ergs of bond energy, which is about 10^{14} atoms. At this rate, the mountain will be wiped away after 10^{30} visits by the old man, that is, after 10^{32} years. The uncertainty in this number is less than two in the exponent, comparable to the uncertainty in the lifetime of a proton.

N8. Olbers's Paradox: Let me continue the reasoning. We start by mentally subdividing the sky about us into evenly spaced concentric spherical shells (of 10 light-years thickness each). The sky contains, on the average, one star in each volume of 100 cubic light-years. By multiplying this density times the corresponding volume, we obtain the number of stars in each shell. In the central sphere (out to a radius of 10 light-years from us) there are 40 stars. In the first shell there are 280; in the second, 760; in the third, 1480 (figure 72). This number increases as the square of the radius of the shell. The stars in each shell are ever farther away, and hence ever fainter to us. Their apparent brightness decreases with the square of the distance. This effect compensates exactly for the increasing number of stars in each shell. As a result, each shell contributes the same amount of light to the night sky. But, in an infinite universe, the number of shells is infinite. This does not mean, however, that the brightness of the sky is infinite. Beyond a certain distance, the surface areas of the stars appear to touch and make a screen. Under these conditions, we obtain a total flux 100,000 times as intense as the Sun's.

N9. The astronomer E. R. Harrison insisted on this point in his book *Cosmology, the Science of the Universe* (London: Cambridge University Press, 1981).

N10. The molecular bond: I would like to describe the nature of this basic force that permits two hydrogen atoms to associate electrically even though each atom is "globally" electrically neutral. (The same sort of phenomenon occurs at the level of complex molecules and plays a fundamental role in the elaboration of organic structures.) Each of the atoms is composed of a proton (positive) and an electron (negative). At a distance, the two atoms ignore each other because of their overall electrical neutrality. But what happens when they approach each other? The positive and negative charges do not have the same distribution in space. The positive charges are points (the nuclei), while the negative charges are spread out in an "electron cloud." The protons repel each other, and the electron clouds repel each other; but each proton attracts the electron cloud of the other atom. As a result, the clouds deform until the attraction and repulsion compensate. The new structure thus formed is a hydrogen molecule. Its bond energy is 4.5 electron volts.

N11. The phenomenon is not simple. There are compensating effects. The reduction, although partially compensated, remains.

N12. See, for instance, L. E. Orgel, *The Origin of Life: Molecules and Natural Selection* (New York: John Wiley, 1973). I also recommend the cassette entitled *Passage de l'inerte au vivant*, by Joël de Rosnay, in the series "Les Après-Midi de France-Culture."

N13. This area bears the name of Isua, an Eskimo word meaning "the farthest you can go."

N14. See, for instance, John Maynard Smith, *The Theory of Evolution* (New York: Penguin, 1976).

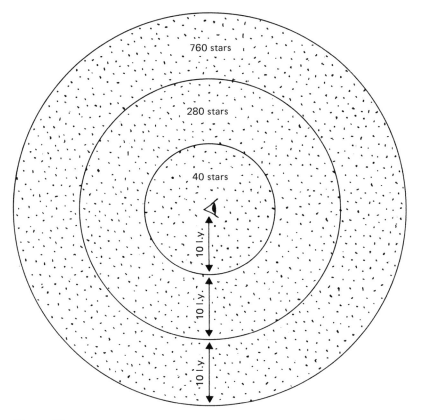

Figure 72
Assuming a constant number of stars per unit volume (per cubic light-year, for example), we find that the number of stars in each shell of 10 light-years thickness increases with distance from the center. This increase compensates exactly for the fact that the more distant stars appear fainter to us.

N15. See Earl Freedon, "The Chemical Elements of Life," *Scientific American*, July 1972.

N16. For a deeper discussion, see I. Prigogine and E. Stengers, *La nouvelle alliance* (Paris: Gallimard, 1980). There is a further complication to the following argument in that left-oriented molecules may spontaneously become right-oriented. The arguments remain quantitatively valid, though.

N17. I am indebted to Pierre Solié for this information and comparison.

N18. For modern physics, the forces arise in the abstract context of gauge theories with local invariance under symmetry transformations. In brief, forces and structures are related to the symmetries of nature. Henri Poincaré was the first to point physics in this direction. It is presently the basis of research in elementary particle physics. We are far from having explored all of its implications.

N19. We then speak of the "degeneracy" of energy levels. In fact, these levels do not need to be strictly identical. It is sufficient that the difference between them be less than the ambient thermal energy (kT).

N20. Here I recommend Stephen Jay Gould's *The Panda's Thumb* (New York: Norton, 1980). Also André Langaney, *Le sexe et l'innovation* (Paris: Seuil, 1979).

N21. I owe this analogy to Richard Dawkins, *The Selfish Gene* (New York: Oxford University Press, 1976).

N22. See J. Barrow and F. Tipler, *The Anthropic Cosmological Principle* (New York: Oxford University Press, 1984).

N23. I mention in this regard some excerpts from a significant article by Alfred Kastler, published in *Le Monde* in May 1977. The article is entitled, "Une machine devenue folle" [A machine gone mad]: "Over the course of the years, the two Great Powers have pursued and intensified militarization at a rate that has now become—and I weigh my words—monstrous and demented. By the term 'monstrous' I mean to pass a moral judgment; by 'demented' I mean to suggest a defiance of human intelligence. Although both nations already possessed, in 1970, nuclear armories sufficient in number for mutual destruction, the two Great Powers have developed, since that date, missiles with multiple warheads called MIRVs. They have granted each other the mutual right, by the treaty of Vladivostok in 1974, to deploy 1320 strategic multiple-warhead intercontinental missiles. A single American Poseidon missile can be armed with ten warheads. The Soviets build missiles with eight warheads. Each of these warheads contains a hydrogen bomb with a destructive power that can range from 40 kilotons up to 2 megatons of TNT, that is, up to 100 times the yield of the Hiroshima bomb (18 kilotons of TNT). In place of a hundred nuclear warheads, the two Great Powers have granted each other the right to make more than 10 thousand each. We know from reliable sources that, at the beginning of 1976, the United States possessed a stockpile of more than a thousand of these engines of megadeath, and the USSR, more than 4 thousand; and they have each likewise built an excessive number of intercontinental missiles capable of being transformed into MIRVs (2400 in the USSR, 2200 in the United States). By virtue of the agreements that were concluded, this nonsensical course is relentlessly pursued."

N24. I refer the reader to the book by Roger Garaudy, *L'Appel aux vivants* (Paris: Seuil, 1979), for further details.

N25. I must add a word for the knowledgeable reader on the famous "curvature" of the universe. In a closed universe, the universe curves back upon itself because of the great density of matter, which makes a "straight line" in space describe a circle. The universe, however, is still unbounded.

N26. The notion of an "elementary particle" is entirely relative to the environment. Everything depends on the amount of energy available. In the cores of stars, atoms dissociate and nucleons become "elementary." In the first microseconds of the universe, quarks were elementary. For a child, television sets are elementary particles, and they are no more to a television repairman equipped with tools.

N27. Laplace, *Essai philosophique sur les probabilités* (Paris: Gauthier-Villars, 1921), p. 3; English edition: *A Philosophical Essay on Probabilities*, translated by F. W. Truscott and F. L. Emory (New York: Dover Publications, 1951).

N28. See especially the discussions of Michel Serres on this subject; for example, in the introduction to *La philosophie première d'Auguste Comte* (Paris: Hermann, 1975).

N29. See the work of Bernard d'Espagnat on this subject: *A la recherche du réel* (Paris: Gauthier-Villars, 1979). For an English summary see *Scientific American*, April 1979.

N30. This is in accord with Carl Jung's idea of synchronicity. See C. G. Jung, *Synchronicity* (London: Routledge and Kegan Paul, 1954).

N31. (Note added in proof, 1983) The data available by late 1983 do not seem to support the view that the neutrino could be massive enough to close the universe.

N32. (Note added in proof, 1983) The data available by late 1983 do not seem to confirm this hypothesis. More precise experiments are being done.

N33. (Note added in proof, 1983) Recent data suggest that planetary nebulae are mostly associated with double stars.

N34. (Note added in proof, 1983) A word of caution needs to be added to this argument. A number of alternatives involving early inflationary phases in the Big Bang have been proposed recently. They could solve the riddle in a perfectly causal way. At present, though, none seems thoroughly satisfactory.

N35. (Note added in proof, 1983) This stage seems to involve mostly double stars and thus may not be relevant to all stars.

Some Numbers to Remember

One light-year: 10 trillion (1×10^{13}) kilometers or 6 trillion (6×10^{12}) miles

Age of the universe: about 15 billion (1.5×10^{10}) years

Age of the Sun: 4.6 billion (4.6×10^{9}) years

Number of stars in a galaxy: approximately 100 billion (10^{11})

Speed of light: 300 thousand (3×10^{5}) kilometers per second or 186 thousand (1.86×10^{5}) miles per second

Further Reading

J. D. Barrow and F. J. Tipler, *The Anthropic Cosmological Principle* (New York: Oxford University Press, 1984)

Niels Bohr, *Atomic Theory and the Description of Nature* (New York: AMS Press, 1976, reprint of 1934 edition)

John T. Bonner, Jr., *The Evolution of Culture in Animals* (Princeton, NJ: Princeton University Press, 1980)

Hanbury Brown, *Man and the Stars* (New York; Oxford University Press, 1978)

The Cambridge Encyclopaedia of Astronomy (New York: Prentice-Hall, 1977)

The Cambridge Photographic Atlas of the Planets (New York: Cambridge University Press, 1982)

Eric Chaisson, *Cosmic Dawn* (Boston: Little Brown, 1981)

Bernard d'Espagnat, "The Quantum Theory and Reality," *Scientific American*, November 1979

René Dubos, *Celebrations of Life* (New York: McGraw-Hill, 1981)

Freeman Dyson, *Disturbing the Universe* (New York: Harper & Row, 1979)

G. B. Field, G. L. Verschuur, and C. Ponamperuma, *Cosmic Evolution: An Introduction to Astronomy* (Boston:Houghton Mifflin, 1978)

Donald Goldsmith, *The Evolving Universe: An Introduction to Astronomy* (Menlo Park, CA: Benjamin-Cummings, 1981)

Stephen Jay Gould, *The Panda's Thumb* (New York: Norton, 1980)

Werner Heisenberg, "Natural Law and the Structure of Matter," in *Across the Frontiers* (New York: Harper & Row, 1974), pp. 104–121

Werner Heisenberg, *Physics and Beyond* (New York: Harper & Row, 1971)

Robert Jastrow, *Red Giants and White Dwarfs* (New York: Norton, 1979)

Robert Jastrow, *Until the Sun Dies* (New York: Norton, 1977)

J. Robert Oppenheimer, *Science and the Common Understanding* (New York: Simon and Schuster, 1954)

L. E. Orgel, *The Origin of Life: Molecules and Natural Selection* (New York: John Wiley, 1973)

Tobias Owen and Donald Goldsmith, *The Search for Life in the Universe* (Menlo Park, CA: Benjamin-Cummings, 1979)

Carl Sagan, *Cosmos* (New York: Random House, 1980)

Joseph Silk, *The Big Bang* (San Francisco: W. H. Freeman, 1980)

Victor F. Weisskopf, *Knowledge and Wonder*, 2d edition (Cambridge, MA: The MIT Press, 1979)

Index